USA HEMP MUSEUM

HEMP FOR VICTORY:

A GLOBAL WARMING SOLUTION

MUSEUM

Hemp Master Richard M. Davis

Photograph by Brenda Kershenbaum

"YOU CANNOT TAKE THE NUMBER ONE
PLANT RESOURCE OUT OF THE ECOSYSTEM
AND EXPECT ANYTHING BUT DISASTER. HEMP MUST BE
RETURNED TO THE PEOPLE FREE AND CLEAR."
HIGH TIMES FREEDOM FIGHTER, FEBRUARY, '95

BY:

RICHARD M. DAVIS

FOUNDER, CURATOR, USA HEMP MUSEUM
A PRIVATE MUSEUM & LIBRARY WITH A VIRTUAL WING

WWW.HEMPMUSEUM.ORG

HEMP FOR VICTORY:
A GLOBAL WARMING SOLUTION

BY: **RICHARD M. DAVIS**

AUTHOR, FOUNDER, CURATOR, USA HEMP MUSEUM

PUBLISHED BY:
HEMPMUSEUM PUBLISHING
A DIVISION OF
THE USA HEMP MUSEUM

A PRIVATE MUSEUM AND LIBRARY WITH A VIRTUAL WING

WWW.HEMPMUSEUM.ORG

WORLD CANNABIS FOUNDATION 501(C)3

FIRST PRINTING, 2007—SECOND PRINTING 2009
ISBN NO: 978-0-9793765-1-1

RESEARCH & SUPPORT:
BRENDA KERSHENBAUM, WORLD CANNABIS FOUNDATION

BOOK COVER DESIGN & ARTWORK:
ORIGINAL "BUY A HEMP FARM" BY MOLLY ENTNER COHEN (1935) –
UPDATE AND ART DIRECTOR SHERWOOD AKUNA (2007)
EDITOR: J. NAYER HARDIN, COMPUTER UNDERGROUND RAILROAD ENT.

PUBLISHED 2007, LOS ANGELES, CALIFORNIA

TO ORDER ADDITIONAL COPIES
WWW.HEMPMUSEUM.ORG

Dedicated to Ed "Ask Ed" Rosenthal

Hemp Author
Hemp Activist
Hemp Publisher
and like the Curator
a convicted
Hemp Felon

WWW.QUICKTRADING.COM

WWW.HIGHTIMES.COM

WWW.CANNABISCULTURE.COM

TABLE OF CONTENTS

HENRY FORD   5

1. HEMP FOR VICTORY: GLOBAL WARMING   6

2. HEMP, BIOFUELS AND GLOBAL WARMING   7

3. HEMP AND GLOBAL WARMING   11

4. HEMP BIO-FUELS AND ENERGY   19

5. AGRICULTURE   36

6. THE WATER FACTOR   62

7. THE POPULATION FACTOR   64

8. THERE WAS ANOTHER EMERGENCY WORLD WAR II   67

9. A NATURAL ENERGY POLICY   76

10. LETTERS TO THE LOS ANGELES TIMES   111

11. READINGS: HEMP AND GLOBAL WARMING   120

12. HEMP LEGISLATION   125

13. 50 THINGS YOU CAN DO TO FIGHT GLOBAL WARMING   152

14. THE U.S.A. HEMP MUSEUM CURATOR'S ROOM   153

APPENDIX   158

Henry Ford

From the Collections of Henry Ford
Museum & Greenfield Village

**"Why use up the forests
which were centuries in the making
and the mines
which required ages to lay down
if we can get the equivalent
of forests and mineral products
from the annual growth of the fields?"**

1. HEMP FOR VICTORY: GLOBAL WARMING
An Introduction To The Series

This is the first book of a series HEMP FOR VICTORY, based on the Museum's extensive collection of hemp information. To solve many of our problems, hemp is a given resource. California, with the nation's sixth largest economy, put legislation on the Governor's desk to grow hemp, but sadly he vetoed the bill. China, England, Russia, Canada, Spain, Italy and many other nations are growing legal hemp.

Because we view global warming as having had a major human contribution, we offer in this first book an immediate workable solution. The world should be in a global survival mode, but nations such as the United States are dragging their feet and withholding cooperation on global warming treaties. Hemp can help.

USA HEMP MUSEUM-1992
CALIFORNIA STATE CAPITOL STEPS, SACRAMENTO, CA

CLEAN HEMP BIO-FUELS CAN COOL
THE 21ST CENTURY GLOBAL WARMING
BY CLEANING THE AIR OF CO2

Visit the virtual hemp museum at www.hempmuseum.org and guess which of the 18 rooms will be next in our series of

HEMP FOR VICTORY

2. HEMP, BIOFUELS AND GLOBAL WARMING

Introduction And Summary

RICHARD M. DAVIS

Photograph by:
Brenda
Kershenbaum

Who: Every person on earth is at risk of dying from the effects of global warming. Global warming is also impacting on food, war, clean water and ultimately, human life on earth. We cannot live and keep our heads in the sand. Hemp biofuel burns clean, and hemp breathes in the excess global warming CO_2 gas from the air as it grows.

Every person on earth deserves the means to food, clean water, shelter and other amenities. Hemp can help at every turn. Our forests are being cut mercilessly, hemp can help. Healthy forests clean the water and restore the carbon dioxide balance. Hemp can supply paper and building materials like press board, plaster, cement and plastics for shelter. Hemp can supply a complete protein and valuable fish type heart healthy oils, without which malnutrition can occur. Hemp is wind pollinated, which helps as an alternative food source in the face of the "disappearing bees" problem.

Hemp biofuels are domestic, plant based energy sources. Hemp grows quickly, breathes in carbon dioxide from the air as it grows, exhales oxygen as it grows, burns clean and can be economically produced and distributed. Wherever petroleum is used, it can be quickly replaced with hemp bio-fuels, which include alcohols, seed oil and wood, to produce energy. Hemp, a safe energy source, can be grown, processed and shipped from the same location without the need for a lot of storage space, empowering family farmers, oil processors and other ancillary businesses.

This shift can empower the true foundation of the American economy, the family farm. AMAZINGLY, WITH HEMP, THE FOSSIL FUELS BURNED AND POLLUTING OUR ATMOSPHERE ARE AVAILABLE ONCE AGAIN AS A RESOURCE, UNTIL A FAVORABLE CO2 LEVEL IS REACHED.

7

What: Hemp bio-fuels have the power to replace fossil based oil, coal and gas quickly. Basic alcohols such as methanol, ethanol or butanol can be made from hemp to power automobiles, trains, planes, and any other power need.

Ethanol is now being used as an additive in gasoline to reduce pollutants. Vegetable oil from hemp can be mixed with methanol to make bio-diesel. Bio-fuels from tree wood are currently being used in thirty-three California energy plants.

Hemp can compete with tree wood, and leave the tree wood as an expanding carbon sink. Rather than coal and other polluting energy sources, alternative energy sources such as hemp must be employed. Hemp fuels do not cause acid rain as do fossil fuels.

When: NOW. THIS IS AN EMERGENCY! Any excuse for not legalizing and subsidizing the growing of hemp to mitigate and reverse global warming puts us all at risk. Until we all rise up and demand government action to legalize the growing of hemp, all we will hear are excuses.

First, convert from petroleum to hemp biofuel and other clean, efficient energies. The emergency in World War II showed us that hemp can be brought into production in as little as two years. Let's get started now. California in 2006 came within one signature of having legal industrial hemp when the Republican Governor Schwarzenegger vetoed the legislation. We hope for a quick reconsideration and passage of that bill.

Where: We need to begin at the source of the problem. Make no mistake about it, we are the problem. We must conserve to show the rest of the world that life can be lived and enjoyed without all the energy we now produce through oil and nuclear. Wherever oil or nuclear energy is used to power things, replace it with a clean, safe alternative like hemp bio-fuel, wind or solar power. We need to research the possibility of using some of our preserved lands to grow hemp on a temporary basis, or subsidize the growing to get it started as Europe has done. Hemp is a strong, sturdy plant that with a little bit of tending, produces great rewards.

Why: Given the dynamics of Global Warming, we are in danger of losing our lives. Tsunami, Katrina and other bizarre weather conditions are clues that when man outlaws nature, and has no respect for her laws, she responds in undesirable ways. Why, because we have a plan of action that has the potential to transform a crisis into a new more self-sufficient and lasting energy policy with hemp.

How: Re-hemp the planet.

Establish a seed bank and empower seed production. Use government land for growing hemp. Ice bergs can be harvested to supply additional irrigation and clean drinking water needs. Issue hemp stamps. Make information available on line to link and empower local growers. Replace farm subsidies with hemp incentives. Replace that which is killing us with that which enriches us.

Diesel trucks and cars can already run on hemp bio-diesel fuel. Other cars can be converted to use any mix of gas or alcohol by using a Flex Fuel Conversion kit. The conversion process creates many short term business opportunities for folks who know how to install a Flex Fuel Conversion kit. On line, have conferences of 'car and biofuel folks' interested in starting businesses to do the conversions. This is a short window (maybe 3-5 years) business, but can generate substantial profits based on volume and fair pricing.

Another promising alcohol fuel from hemp biomass is butanol. "According to Environmental Energy, Inc., butanol can run in unmodified gas cars. In the summer of 2005, EEI drove an unmodified '92 Buick across the US running on butanol...EEI uses a patented, two-stage process to convert biomass into butanol." (www.solarpower.org)

Let each individual begin within.

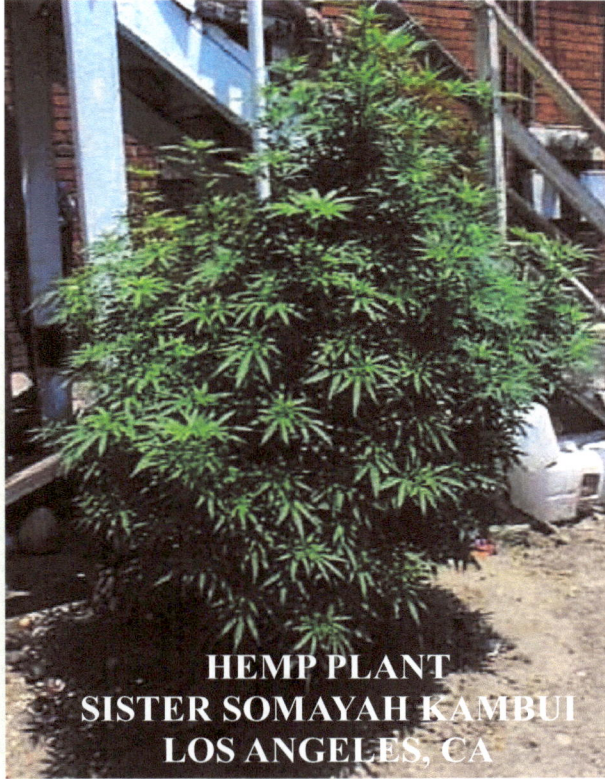

HEMP PLANT
SISTER SOMAYAH KAMBUI
LOS ANGELES, CA

Learn hemp

Buy hemp!

Think hemp!

Eat hemp!

Grow hemp!

Enjoy hemp!

HEMP HERO GEORGE WASHINGTON

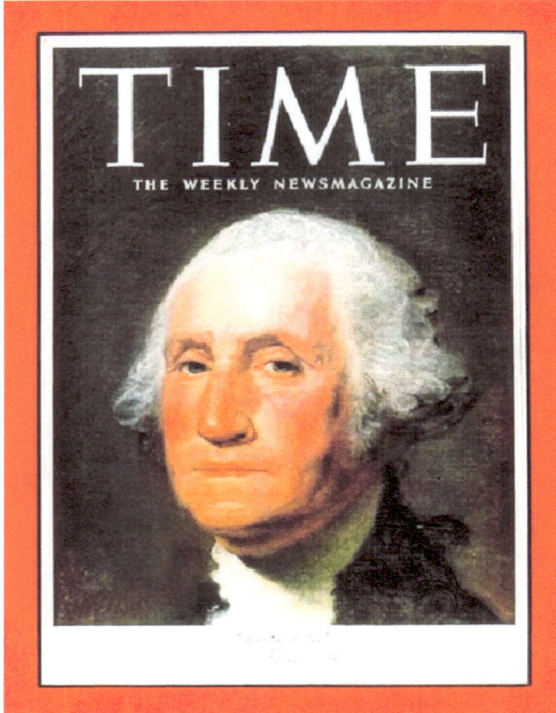

FATHER OF THE U.S.A.,
FIRST PRESIDENT,
REVOLUTIONARY
WAR GENERAL
AND HEMP FARMER,
GEORGE WASHINGTON
SAID:

"MAKE THE MOST OF THE
INDIAN HEMP SEED,
SOW IT EVERYWHERE."

GEORGE WASHINGTON
AND HIS SLAVES
GREW HEMP AT HIS
MT. VERNON HOME
IN VIRGINIA.

3.HEMP AND GLOBAL WARMING

"THE ERA OF PROCRASTINATION,
OF HALF-MEASURES, OF SOOTHING AND BAFFLING
EXPEDIENTS, OF DELAYS, IS COMING TO ITS CLOSE.

IN ITS PLACE WE ARE ENTERING
A PERIOD OF CONSEQUENCES."

WINSTON CHURCHILL, 1936

Any discussion on Global Warming must realize that what is involved is nothing less than survival of the planet as we know it. As Churchill said, we have entered a period of consequences. Drastic changes in how we operate society on a planetary level will be necessary to cool the earth after a century of fossil fuel madness and no thought for future generations.

This book offers some hope that we can stem the tide of warming and its consequences using among other measures the hemp plant. There was another emergency where hemp prohibition was lifted and hemp planted to save the world – see Chapter 8.

I often said to myself if I ever write a book I would have to dedicate it to the people of the U.S. for helping finance my lengthy college education with student loans, the G.I. Bill, a U.S. Public Health Service Fellowship. Over that ten year period I rarely missed class and achieved an incredible amount of personal growth. Thank you.

In a real way this is a report of two experiments in the use of Cannabis sativa, hemp, or marijuana. The first experiment, to smoke hemp daily for thirty years was started at Cal. State Los Angeles where I received a B.S. in Zoology and a Masters in Biology, and was continued at the School of Public Health at U.C.L.A. I was close to getting a doctorate, and had been smoking pot daily for six years. Studies were coming through with reports of brain damage, and other maladies due to marijuana that could not stand any scientific review.

Junk science, scare tactics, and plain lies seemed to be the order of the day for marijuana studies. Here was something I had experience with and was passionate about. My advisor wouldn't let me do my doctoral thesis on marijuana so I dropped out and moved to Northern California where I could grow my experimental stash.

The second experiment started when, one day, I found out about hemp's potential as a resource crop. I had been smoking pot for 22 years by then. People were talking about hemp, I was growing it. I knew first hand how fast hemp grows, I was a hemp farmer. I learned George Washington, Thomas Jefferson, and my Mother's ancestors in Missouri grew hemp. I had the resource! I learned to make paper, boxes, boards, felt, plastic, varnish, twine, yarn, rope, cement, and with a dozen ways to make hemp medicines I started what has become the U.S.A. Hemp Museum now on the internet at www.hempmuseum.org. (See chapter 14)

HEMP HERO THOMAS JEFFERSON

The fourth U.S.A. President, Statesman, author of the Declaration of Independence, hemp farmer, and slave owner Thomas Jefferson said:

"THE GREATEST SERVICE WHICH CAN BE RENDERED ANY COUNTRY IS TO ADD A USEFUL PLANT TO ITS CULTURE."

America's First Gardener

DIANE ACKERMAN

Jefferson traded seeds, cuttings and botanical observations with naturalists worldwide.

Today, he no doubt would be an environmental activist, probably denounced by some as a "tree-hugger."

Article by Diane Ackerman, PARADE MAGAZINE, July 15, 2001.

"Gardening is a favorite activity in more than 50 million American households, but for THOMAS JEFFERSON -who, in his spare time, was President -it was an all consuming passion...As a slaveholder, he had an almost endless supply of labor for the task."

Hemp

Photo by Ed Rosenthal

The above cover of HEMPWORLD: The International Hemp Journal for Spring 1996 shows Ben Dronkers, the most dangerous man in hemp, at a field in northern Holland. Mari Kane, Editor/Publisher

What is hemp? Hemp is the most important plant in the history of humans and until recently one of the least known in the U.S. The history of hemp dates to prehistoric man's first attempt at agriculture for fiber production, some ten to fifteen thousand years ago. However, wild hemp was available to other hominid lines when they migrated out of Africa millions of years ago into the Middle East. Our present species of man, *Homo sapiens*, arose a mere 100,000 years ago in Africa, so it is likely our distant relatives spread hemp to the rest of Asia and Africa.

Humans evolved and became civilized with hemp at their side. The first and oldest use of hemp was no doubt for food, the seed being a complete protein and laden with health giving essential oils. The oldest known fabric is hemp. The oldest known true pulp paper is of hemp; hemp ropes and sails carried the commerce of the world for some 6,000 years of sailing the seas. Hemp symbols in writing are 5,000 years old, as are hemp medicines. Hemp is one of the plants that made civilization possible for human animals. And in the United States of America its citizens are prohibited from growing hemp. This is absurd and dangerous thinking when we are all confronted by rising atmospheric temperature. Hemp can help.

It was estimated by Popular Mechanics Magazine, 1938, that there were 25,000 viable uses for the hemp plant (now estimated at 50,000 uses). It was touted as the new billion dollar crop by the Magazine.

Plants, which dominate life on land in terms of volume, absorb around 102 Gt (1Gt = 1 billion tonnes) of carbon per year which is drawn down during photosynthesis – the production of organic molecules from carbon dioxide and water in the presence of sunlight (*Global Warming*, Greenpeace, 1990, p.24).

The normal non-human carbon cycles of land and ocean are in equilibrium. We are pulling age old carbon called fossil fuel, coal, crude oil, and natural gas long buried out of the ground and burning it in our factories, in our power plants, in our homes and cars – and have for the past 200 years.

15

"Global climate change is the most threatening and intractable of all environmental problems we face. Carbon dioxide (CO_2) is the most crucial of the greenhouse gases contributing to global warming. Since pre-industrial times, CO_2 levels have risen by almost 30% due to deforestation and fossil fuel combustion. The U.S. currently burns fossil fuels for 93% of its energy needs and consumes 25% of the world's supply. One tank of gasoline generates up to 400 pounds of CO_2." (Lee Hitchcox, D.C., HEMP WORLD, Vol. 4, Number 1, 1998)

CELLULOSE ATOMIC STRUCTURE

The structure of cellulose above shows the number of carbon atoms (11) in each molecule. Plants breathe in carbon dioxide (CO_2 is the primary greenhouse gas because of its volume) and breathe out oxygen (O_2).

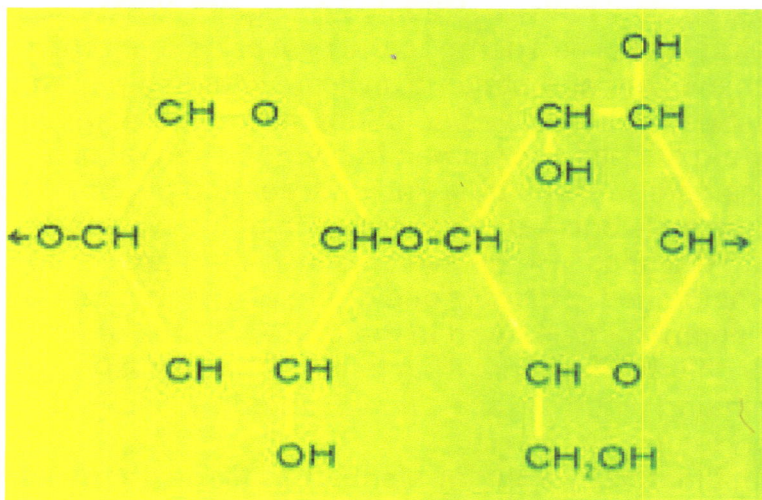

"Gas exchange in biology is the exchange of gases between living organisms and the atmosphere, principally oxygen and carbon dioxide. In animals, gas exchange is only respiratory (or using oxygen to convert food to energy). In plants, gas exchange is photosynthetic (or using carbon dioxide to make food) as well as respiratory... In plants, gas exchange necessary for photosynthesis and respiration generally takes place via the stomata, small pores in the above ground parts of a plant, and on the undersurface of leaves where there may be as many as 300,000 per square inch." –Webster's New World Encyclopedia, 1992.

**"...THE EARTH IS SLOWLY DYING,
AND THE INCONCEIVABLE -- THE END OF LIFE ITSELF –
IS ACTUALLY BECOMING CONCEIVABLE.
WE HUMAN BEINGS OURSELVES
HAVE BECOME A THREAT TO OUR PLANET..."**

**Queen Beatrix of the Netherlands
in her Christmas message to the people of Holland, 1998**

Global Warming

What is global warming? The planet is being warmed by what is called the greenhouse effect. Excess atmospheric gases block radiant heat from escaping back into space in much the same way a greenhouse or closed car heats up in the sun.

"Global warming represents a disruption of the environment that is without parallel in human experience... The most abundant greenhouse gas is carbon dioxide (about 55% of warming). Other atmospheric gases that are causing global warming are methane (15% of warming), CFC's (24% of warming) and nitrous oxide (6% of warming) [1980's figures, pps. 17 & 97, *Global Warming*].

What are the sources of these greenhouse gases? The main source of human generated carbon comes from fossil fuels –coal, oil, and natural gas, from pre-existing life. Burial, compacting, and heat changed this pre-existing life carbon into the fossil fuels we know today, previously cut off from the natural carbon cycle millions of years ago.

"In a world rapidly becoming inured to sweeping change on the political stage, we have witnessed the emergence of an environmental threat which cuts to the heart of how humans choose to operate society – a problem which is truly global in both consequences and cause. Greenhouse gases are produced in their current superabundance as a result of the ways we humans produce and use energy, by the use of certain industrial chemicals (CFCs and related gases), and by intensive agriculture and tropical deforestation. In a world in which the greenhouse effect is allowed to continue its buildup, we would all – at some stage- be losers, and we would all – to varying degrees – be responsible." (*Global Warming*, Greenpeace, 1990, p.2.)

17

Global warming has happened before. A recent newspaper article in the *Los Angeles Daily News*, August 26, 2006, titled: "It's happened before: Clues to global warming." The discovered ancient hot spell, called the Paleocene-Eocene Thermal Maximum (PETM), occurred 55 million years ago, and lasted 50,000-100,000 years. Earth's temperature rose 10 -12 degrees as carbon was dumped into the atmosphere by some geological catastrophe over 10,000 years to start the PETM period. This period led to the extinction and relocation of many plants and animals of both land and ocean.

Today's warming is happening on a much faster scale than PETM. What took 10,000 years to build up enough carbon to start PETM; we could do in 500 years if it's business as usual until 2300 A.D.

Today's warming is evident the planet wide. Al Gore's movie and book, *An Inconvenient Truth*, 2006, shows dry lakes, melting glaciers, and other climatic changes which are *evidence of drastic changes underway now*. People interested in the survival of the planet should re-elect Al Gore to the U.S. Presidency.

One 'stabilizing scenario' from *Halting Global Warming* by Mick Kelly (from *GLOBAL WARMING*, 1990, Greenpeace, supra, p. 105), consists of the following goals:

1. The elimination of the production of chlorofluorocarbons and all related ozone-depleting chemicals by the year 1995 and the avoidance of substitutes that are greenhouse gases;

2. A halt to deforestation, followed by extensive reforestation to offset 1.65 Pg (1,650 million metric tons) a year of energy-related carbon emissions by the year 2020;

3. A reduction in carbon emissions from fossil-fuel combustion; and

4. A reduction in the annual rise in methane and nitrous oxide concentrations to 25% of the present value by 2020.

These are global goals that require global understanding of the crisis and global cooperation which does not exist in 2006. The United States, with 5% of the world's population produces 25% of the world's greenhouse gases, has refused to join the 1997, Kyoto Treaty whose goal is to control global warming pollution. We have an answer in hemp that is being ignored by leaders who would rather go to war over oil than admit losing to hemp advocates. The reality is that we can now grow hemp (marijuana) for medicine in California and ten other states. Why not for world health?

Why not for our own health?

18

4. HEMP BIO-FUELS AND ENERGY

Above, hemp hurds or hemp wood
sun bleached after the fiber's matrix was retted
(rotted) by two weeks in water and the fiber
was stripped from the wood.

Hemp hurds are 77% cellulose,
including some excess greenhouse carbon.

"Anything you can make from hydrocarbons
(oil, coal, natural gas),
you can make from carbohydrates
(plant material)."

DR. WILLIAM HALE

Hale argued that farms not oil wells
should be the source of biochemicals,
and warned of petrochemical
pollution problems on health.

[THANKS TO HAUPTLING ABERJA – email respondent]

Biofuel Uses

AUTOMOBILES
TRUCKS
TRAINS
PLANES
MOTOR CYCLES
SHIPS AND BOATS
SMALL ENGINE USES
POWER PLANTS
PEACE TOOL (NO WARS FOR OIL)
REDUCE CITY SMOG
REDUCE ACID RAIN
MITIGATE GLOBAL WARMING

HEMP CLEANS CO2 FROM THE AIR AS IT GROWS

HEMP BURNS CLEAN AS IT PROVIDES ENERGY

PATCH FROM
WWW.HEMPTODAY.COM

"CORVALLIS, Ore. 2007 – A new economic analysis of biofuels by Oregon State University sets a cautionary tone for the large-scale production of biofuels in Oregon. Results of the study suggest that the "net energy" of biofuels is expensive when all costs of its production and delivery are taken into account.

The study was released this week by a team of economists in OSU's College of Agricultural Sciences that included William Jaeger, Robin Cross and Thorsten Egelkraut.

By subtracting the energy spent to produce raw materials and to process and transport the biofuels, the researchers found that the cost of the net gain in energy for these biofuels may be more than seven times higher in some cases when compared to gasoline.

There is a commercial market for biofuels in Oregon given current subsidies," Jaeger said. "But success in the marketplace doesn't mean cost-effectiveness in achieving the state's goals of energy independence and reducing greenhouse emissions."

The study was prompted by increasing interest in domestically grown biofuels as an alternative to foreign imports of oil. The economists examined three biofuels options for Oregon: ethanol made from corn, ethanol made from wood cellulose, and biodiesel made from canola.

For each option, the researchers examined the cost of production, its contribution to energy independence and its environmental impact in terms of greenhouse gas emissions. They calculated "net energy" as the amount of energy in the biofuels minus the amount of energy it takes to produce, process, and transport the biofuels.

Their results suggest that **ethanol made from wood cellulose produced the greatest net energy, netting 84 percent** of its energy after production fuel costs were subtracted. Biodiesel made from canola netted 69 percent of its energy after subtracting production fuel costs. And **ethanol made from corn netted a mere 20 percent** of its energy after subtracting the energy spent to produce it."

STOP BURNING FOOD FOR FUEL

HEMP IS FOUR TIMES MORE EFFICIENT THAN CORN AS BIOFUEL

HEMP BURNS CLEAN
PRODUCING CLEAN ENERGY

The U.S.A. Hemp Museum has been interested in fuels and energy from its start in 1990. As the Hemp Museum's Curator, I spent many hours in the library of the California Energy Commission (C.E.C.) researching biomass (plant matter) for fuels and energy. During the 1992 election campaign I wrote an article called *A NATURAL ENERGY POLICY*, which is included in this book. Among the things I learned from the C.E.C. was that Sacramento had a power plant not far from the Capitol built to burn biomass collected as waste tree, shrub, and grass trimmings. The plant was not in operation. As it was explained to me the green matter to be used as fuel was always too wet and irregular in composition to adequately fire the power plant. California now has 33 biomass power plants in the state, operating mostly on forest logging waste. These power plants could burn year round with hemp for energy.

Hemp will produce cleaner air and reduce greenhouse gases. When biomass fuel burns, it produces CO_2 (the major cause of the greenhouse effect), the same as fossil fuel; but *during the growth cycle of the plant, photosynthesis removes as much CO_2 from the air as burning the biomass adds, so hemp actually cleans the atmosphere*. With the first cycle there is no further loading to the atmosphere.

If the hemp is not burned, but used in textiles, paper production, building materials, plastics, or other uses then this recycled CO_2 is not released back into the atmosphere and global warming is slowed and halted.

The U.S. must live up to the international treaty to reduce greenhouse gases to 1990 levels. Hemp is the change voted for in America's historic 2008 election, winning in 9 out of ten states. When hemp biomass is used for other more permanent applications, say a library book that will last 1500 years, or building materials in a home (I never thought what it might do to the price of a home), or plastics, or textiles, potential greenhouse carbon is tied up and does not go back into the atmosphere. Hemp can be recycled seven times. Tell elected and appointed officials to use hemp to empower our lives.

AMAZINGLY, WITH HEMP, THE FOSSIL FUELS BURNED AND POLLUTING OUR ATMOSPHERE ARE AVAILABLE ONCE AGAIN AS A RESOURCE, UNTIL A FAVORABLE CO_2 LEVEL IS REACHED.

HEMP BURNS

Tree wood pellet pulp, hemp wood pellet pulp and hemp wood pellets, hand made in the early 1990's for use in pellet burning wood stoves as a demonstration for the Hemp Museum. In 2006 a Canadian company wrote me of their plans to market such a hemp pellet. Europe and Canada where "industrial hemp" is grown and subsidized have the jump on the United States in the development of the hemp culture and products to come to the world out of necessity and for survival.

Quest Hemp Pellets

Industrial Oil
Item #Seed-06

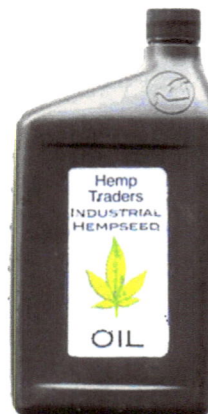

Description: The oil pressed from the hemp seed which is not processed for food is referred to as "Industrial Hemp Seed Oil". This oil has many uses, most often being used as body oil for soaps, lotions, lip balm, and shampoo. But that is only the beginning. Hemp seed oil has a long history of industrial applications. Hemp was the perfect source for a natural based drying oil. For thousands of years, virtually all good paints and varnishes were made with hemp seed oil. Hemp seed oil was also widely used as lighting oil, up until the late 19th century. It burns evenly and is even capable of powering a diesel engine. Use hemp seed oil and say goodbye to petroleum! The long chain saturated and unsaturated fatty acids which constitute the oil make an excellent chemical precursor to other industrial compounds.

Distribution Channel	Quantity/Weight	Price
Sample/Retail	1 Liter +	$16/Liter
Wholesale	20 Liters	$8/Liter
Wholesale	100 Liters	$6/Liter

Fortune Magazine

September 19. 2005 Special Issue - 75th Anniversary

HOW THE WORLD WILL WORK,
THE NEXT 75 YEARS: ENERGY, MIDDLE EAST

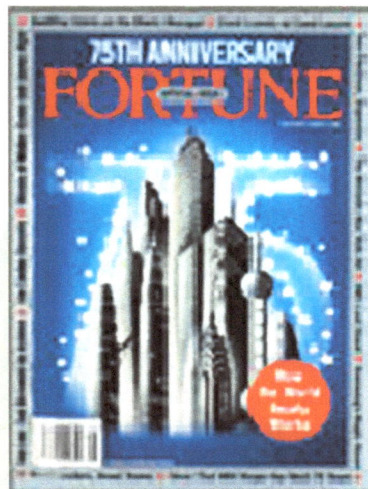

"Weary of all the drawbacks associated with fossil fuels, the world makes a concerted effort to kick the habit. In 2040 oil consumption begins to fall in absolute terms, and by 2060 oil is a boutique fuel. Oil-dependent Middle East economies, which had never diversified, take a brutal hit—sparking violence. But then a promising generation of reformers emerges to replace their blundering predecessors. Their stated mission" devising a freer political system.

Iowa. Pushed by the effects of 2030's oil shock, the alchemy of turning crops into energy is finally mastered. Bio-fuels become big business, and young entrepreneurs flock to the heartland.

Singapore is the first country to ban non-hydrogen cars. By 2050, hybrid hydrogen-electric vehicles are king of the road everywhere. Tooling around in an oil-fueled SUV is regard with as much horror as clubbing, micro turbines fueled by a variety of sources provide on-the-spot power.

Africa. One big beneficiary: Africa. No longer reliant on corrupt politicians to extend the grid, many communities finally have reliable power. Other than a handful of new nuclear plants, large-scale power projects are rare. Instead, micro turbines fueled by a variety of sources provide on-the-spot power. And the civic organizations that made it happen gain force. These two trends bring new spirit to the continent."

IF THE THOUGHT AND TECHNOLOGY ARE IMMINENT,
WHY NOT NOW!

GLOBAL WARMING IS SERIOUS NOW,
AND THE TIME TO REACT TO THESE CHANGES
MAY BE VERY SHORT.

WE DO NOT KNOW.

As a direct fuel, Hemp stalks continue to fuel the wok cooking of much of China (right). Hemp seed was eaten by humans before recorded time, and hemp seed oil was used as lamp oil throughout history. Hemp or Cannabis medicines have a 5,000 year history of safe use, with not a single death recorded.

Hemp in history was burned as oil in lamps for light, and it is much less smelly than kerosene based lamp oil. Oil from hemp seeds can power existing diesel engines (see sample of bio-diesel below), with reduced sulfur and carbon monoxide emissions. Bio-mass (vegetation or plant matter) fuels such as methanol (wood alcohol) or ethanol (grain alcohol) can power modified gasoline engines, or supply hydrogen for fuel cell applications. Both can be made from hemp. Twenty thousand methanol burning cars were tested in California on government and utility fleets. Seventy methanol pumps were installed around the state. Hemp stalks are the best biomass on the planet. And again because it is so important, bio-mass crops absorb carbon dioxide emitted by cars and power plants, mitigating the greenhouse effect.

Hemp car was an alternative-fuel project car that utilized hemp biodiesel for fuel. Industrial hemp would be an economical fuel if hemp were legal to cultivate in the United States. Industrial hemp has no psychoactive properties and is not a drug. Hemp Car demonstrates the concept of hemp fuels on a national level and promotes the reformation of current law.

WWW.HEMPCAR.ORG

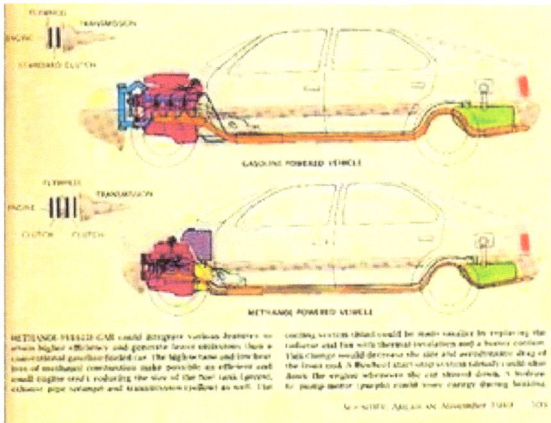

METHANOL-FUELED CAR could integrate various features to attain higher efficiency and generate fewer emissions than a conventional gasoline-fueled car. The high-octane and low-heat loss of methanol combustion make possible an efficient and small engine and t reducing the size of the fuel tank (green), exhaust pipe (orange) and transmission (yellow) as well. The cooling system (blue) could be made smaller by replacing the radiator and fan with thermal insulation and a heater core. This change would decrease the size and aerodynamic drag of the front end. A flywheel start-stop system (detail) could shut down the engine whenever the car slowed down. A hydrogen-to-motor (purple) could store energy during braking.

SCIENTIFIC AMERICAN November 1989

SCIENTIFIC AMERICAN
November, 1989
Article:
"The Case for Methanol"

Methanol-fueled car could integrate various features to attain higher efficiency and generate fewer emissions than a conventional gasoline-fueled car. Cool burning methanol needs no radiator so the front of the car can be very streamlined.

Methanol and electric,
with flywheel design,
the earth would love it.

**Alternative Energy Conference:
Liquid Fuels, Lubricants
and Additives from Biomass.**

The U.S.A. Hemp Museum
invited itself to the
Alternative Energy Conference
whose program is shown here
(right) held at the Weston Crown
Center, Kansas City, Missouri,
June 15-18, 1994.

The University of Iowa had
driven a 3/4 ton bio-diesel
powered truck to the
conference, carrying
its own fuel.

The bio-diesel was made from
the oil of rapeseed (or canola)
and 10% methanol. Samples of
the rapeseed and bio-diesel
were donated to the museum.

The museum was set up
just outside the conference
room, and was very well
received by the
energy group attending.

Preliminary

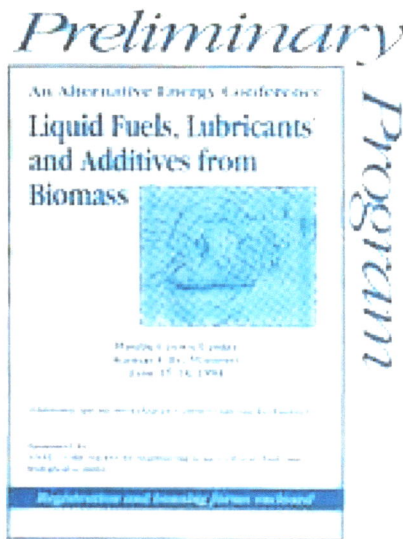

An Alternative Energy Conference
Liquid Fuels, Lubricants
and Additives from
Biomass

Program

This is how to meet the requirements of treaties on global warming:

Plant hemp for biomass and use it as a resource in place of fossil fuels.

We have known this for fifteen years and have done little about it.

"If coal-fed methanol plants were required to keep carbon dioxide emissions to a minimum, and if methanol vehicles were made to be highly efficient, it is projected that total carbon dioxide emissions might be as little as one-fifth of the amount now generated by vehicles that burn gasoline refined from crude oil. If methanol were extracted from biomass, which absorbs as much carbon dioxide as it emits, no further carbon dioxide would enter the atmosphere, and contributions to global warming would be negligible."

Reprinted, this is part of the previously mentioned article:

SCIENTIFIC AMERICAN, November, 1989. Article:

"The Case for Methanol"

These are the words of the above two Environmental Protection Agency scientists:

1992 LUMINA VARIABLE FUEL VEHICLE

This 1992 Lumina Variable Fuel Vehicle runs on any mix of gasoline and alcohol that you put in the tank. I don't know if they ever tried straight vodka.

Twenty thousand flex fuel methanol burning cars were tested in California on government and utility fleets.

Seventy methanol pumps were installed around the state in 1992.

1992 LUMINA VARIABLE FUEL VEHICLE

ORDERING INFORMATION

Methanol

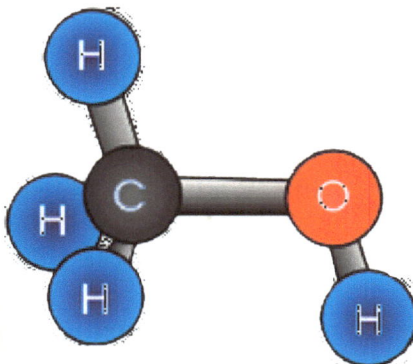

METHANOL (common name methyl alcohol) CH30H is the simplest of the alcohols. It can be made by the dry distillation of wood [like trees or hemp, hence it is also known as wood alcohol], but it is usually made these days from coal or natural gas.

When pure, methanol is a colorless, flammable liquid with a pleasant odor, and is highly poisonous and corrosive. Methanol is used as a chemical feedstock. See Hemp Museum Chemical Feed Stocks Room.

Hemp is the number one source of biomass for making ethanol or methanol (also known as wood alcohol) for fuel. As stated by the Environmental Protection Agency (EPA) of the federal government, **biomass is the cheapest source of energy known** (*SCIENTIFIC AMERICAN*, November, 1989. Article: "The Case for Methanol.")

METHANOL
A clean fuel leader!
DRIVE CLEAN CALIFORNIA

Ethanol

ETHANOL: From: WESTWAYS, AAA Magazine, March/April 2006, KNOW YOUR WHEELS SECTION:

"Short for ethyl alcohol (also known as grain alcohol), ethanol is distilled from biomass sources such as sugarcane, corn, and algae. It's mixed with gasoline as a way to reduce emissions, boost octane, and extend fuel supplies. Ethanol contains less heat energy than gasoline; thus, the mixture's miles-per-gallon rating is reduced."

And again because it is so important, bio-mass crops absorb carbon dioxide emitted by cars and power plants, mitigating the greenhouse effect.

By most accounts booze is better burned in the car than in the body. And while it may not be the only answer, it may be one of the answers to future energy needs and can be made from hemp with the help of enzymes.

Ethanol, common name ethyl alcohol C_2H_5OH, is the alcohol found in beer, wine, cider, spirits, and other alcoholic drinks. When pure, it is a colorless liquid with a pleasant odor, miscible with water or ether, and which burns in air with a pale blue flame. The vapor forms an explosive mixture with air and may be used in high-compression internal combustion engines. It is produced naturally by the fermentation of carbohydrates by yeast cells. Industrially, it can be made by absorption of ethene and subsequent reaction with water, or by the reduction of ethanol in the presence of a catalyst, and is widely used as a solvent. Ethanol has emerged as an additive to gasoline to reduce emissions in high smog states like California.

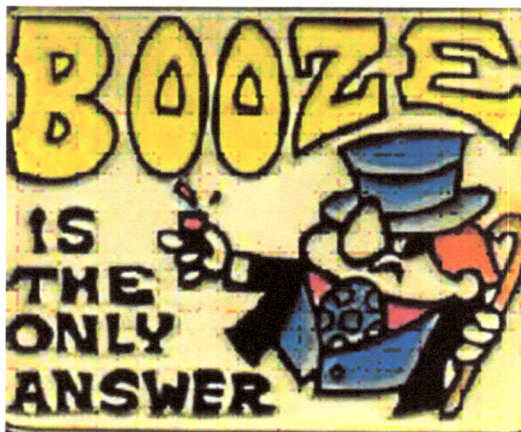

30

Butanol

Another promising fuel is bio-butanol. "According to Environmental Energy, Inc., butanol can run in unmodified gas cars. In the summer of 2005, EEI drove an unmodified '92 Buick across the US running on butanol…EEI uses a patented, two-stage process to convert biomass into butanol." (www.solarpower.org)

"Although ethanol and butanol are both alcohols, butanol has a higher energy content, can be more easily blended with gasoline, and can blend at higher concentrations without retrofitting cars…Bio-fuels now account for 4% of U.S. gasoline consumption. What that means…is the U.S. must raise annual production of bio-ethanol from 5 billion gal currently to 60 billion within 25 years.

"Eventually, most ethanol plants will have to run on non-edible, cellulosic feed stocks [of which hemp is king], rather than the corn that fuels them today. The U.S. cannot produce nearly enough corn to supply food needs and still meet ethanol targets."

Most of the hemp products made by the Curator started as cut-up small branches such as are in the bottle second from the right above. On the right is rough blended pulp for boards. Left bottle contains finer pulp for paper and pellet stoves. Second from left bottle contains fine pulp from tree wood pellets for stoves.

Biodiesel

In 1892, Rudolph Diesel invented the diesel engine, which he intended to fuel "by a variety of fuels, especially vegetable and seed oils."

Biodiesel fuel may be mixed in any ratio with petroleum diesel. Dynamometer tests indicate full power output with up to 75% reduction in soot and particles. No engine modification is needed to burn bio diesel fuel.

According to Aqua Das, biodiesel from hemp could be the fuel of the future.

"Vegetable Oil Will Fuel New Jersey Test Buses!

(Source Inv. Buss. Daily)

NJ Transit is conducting a four-month test of a blended diesel fuel containing vegetable oil, like that from soy beans (hemp-seed). The National Bio Diesel Board is providing the fuel free of charge, thanks to a grant from the United Soybean Board. The B-20 fuel, containing 20% vegetable oil, was developed by Twin Rivers Technologies, Inc. of Quincy, Mass. "(1997)

Direct Biomass Fuel

But can't we just burn the hemp?   Yes and the power plants are here.

Hemp can produce electricity. There are 33 biomass power plants in California now capable of burning hemp for electrical power. These plants now burn forest wastes from logging that must be stopped. And these wastes are not available year round due to weather in the forests. Hemp can do it.

Flex Fuel Conversion Kits

One of the ways to convert a standard engine to one that can run on hemp ethanol is a flex fuel conversion kit. They run between $400-$1,000, but with that, one can be off of oil and onto hemp fuel. If liquid bio-butanol works out as a fuel, it may be that no conversion is necessary, saving billions of conversion dollars.

A benefit of the conversion process is a large number of unemployed auto workers can find work, or set up their own companies, doing the conversions from gasoline to bio-fuel.

Flex Fuel System

Diagram From

WWW.ABOUT.COM

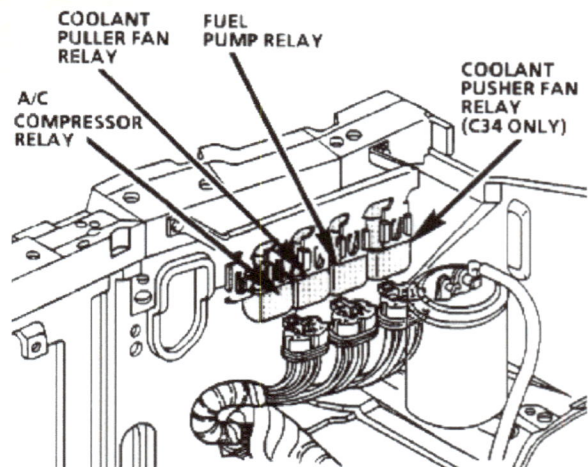

32

Energy & the Economy

From <u>The Emperor Wears No Clothes</u> By

HEMP HERO JACK HERER

"The book Solar Gas, Science Digest, Omni Magazine, The Alliance for Survival, the Green Party of Germany, the United States and others put the total figure of our energy costs at 80% of the total dollar expense of living for each human being."

WWW.JACKHERER.COM

In validation, 82% of the total value of all issues traded on the New York Stock Exchange and other world stock exchanges, etc., are tied directly to:

Energy producers such as Exxon, Shell Oil, Conoco, Con-Edison, and so forth.

Energy transporters such as pipeline companies, oil shipping and delivery companies.

Refineries & retail sales of Exxon, Mobil, Shell, So. California Edison, Con-Edison, etc.

Eighty-two percent of all your money means that roughly 33 of every 40 hours you work goes to pay for the ultimate energy cost in the goods and services you purchase, including transportation, heating, cooking, lighting. Americans - 5% of world population - in our insatiable drive for greater "net worth" and "productivity," use 25-40% of the worlds' energy. The hidden cost to the environment cannot be measured.

Our current fossil energy sources also supply about 80% of the solid and airborne pollution which is quickly poisoning the environment of the planet. (See U.S. EPA report 1983-96 on the coming world catastrophe from carbon dioxide imbalance caused by burning fossil fuels). The best and cheapest substitute for these expensive and wasteful energy methods is not wind or solar panels, nuclear, geothermal and the like, but the evenly distributed light of the sun for growing biomass.

On a global scale, the plant that produces the most net biomass is hemp. It's the only annually renewable plant on Earth able to replace all fossil fuels.

In the Twenties, the early oil barons such as Rockefeller of Standard Oil, Rothschild of Shell, etc., became paranoically aware of the possibilities of Henry Ford's vision of cheap methanol fuel,* and they kept oil prices incredibly low - between one dollar and four dollars per barrel (there are 42 gallons in an oil barrel) until 1970 - almost 50 years! Prices were so low, in fact that no other energy source could compete with it. Then, once they were finally sure of the lack of competition, the price of oil jumped to almost $40 per barrel over the next 10 years.

* Henry Ford grew marijuana on his estate after 1937, possibly to prove the cheapness of methanol production at Iron Mountain. He made plastic cars with wheat straw, hemp and sisal. (Popular Mechanics, Dec. 1941, "Pinch Hitters for Defense.") In 1892, Rudolph Diesel invented the diesel engine, which he intended to fuel "by a variety of fuels, especially vegetable and seed oils."

What's In The Way Of Re-hemping Now?

THE HEMP LEAF
WAS GIVEN A BAD NAME
BY THE U.S. GOVERNMENT IN THE
MARIJUANA TAX ACT OF 1937,
WHICH TURNED INTO A FULL PROHIBITION.

5. AGRICULTURE

Richard M. Davis with hemp crops

HOW TO FINANCE?

TAXES WOULD BE PAID ON HEMP

A 20% tax on recreational hemp could support the healing of Global Warming, by abundantly financing our agricultural needs, as we navigate these interesting times. This income could also help folks deal with environmental disasters, i.e. Hurricane Katrina.

This hemp tax fund would empower the style of a real "Uncle Sam," a person folks can turn to for help and constructive advice in times of need.

HEMP'S US AGRICULTURAL HISTORY

FIRST SEEDS

Pictured right is an image of the Mayflower II built in 1957 and equipped with hempen cordage. (National Geographics) Early Settlers from Europe brought hempseed to the Americas. England wanted another colony with the ability to grow hemp for their massive navy, but farmers grew the more profitable crop -tobacco. Much of the hemp grown was used locally as homespun fabric and rope.

37

George Washington was great in so many ways, it has been forgotten that he was a hemp farmer and advocate of hemp industry. Washington said,

"Make the most of the hempseed and sow it everywhere."

HEMP FARMER & U.S. PRESIDENT

GEORGE WASHINGTON

U.S. President Thomas Jefferson was a hemp farmer

"Jefferson's annual goal was 1,200 yards of cloth woven from purchased cotton and wool *and hemp produced on his farms*."

WWW.MONTICELLO.ORG

It is ironic that the prohibited hemp plant produced the material for both the early American flags and the paper upon which our Supreme Law is written-the U.S. Constitution.

The Hemp Plant, male (left) and female (right) and Seed.

**This Hemp Museum print of a hand-painted
botanical illustration was made when hemp was a free plant.**

**IN 1938, IT WAS ESTIMATED
BY POPULAR MECHANIC MAGAZINE THAT
THERE WERE 25,000 VIABLE COMMERCIAL PRODUCTS
POSSIBLE FROM THE HEMP PLANT.
TODAY IT'S OVER 50,000 USES FOR THE HEMP PLANT.**

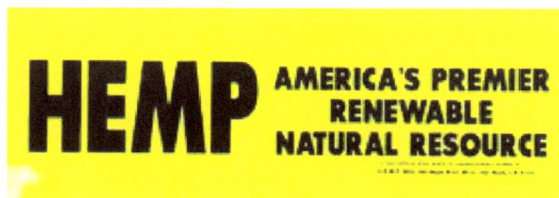

HEMP AMERICA'S PREMIER
RENEWABLE
NATURAL RESOURCE

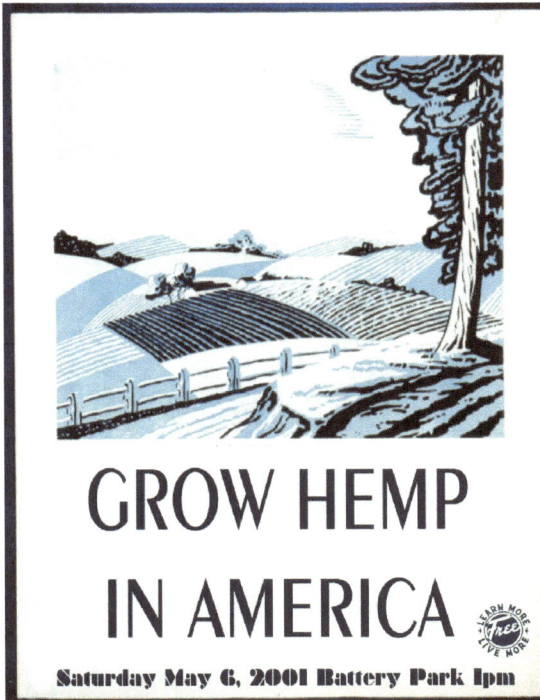

GROW HEMP IN AMERICA

Saturday May 6, 2001 Battery Park 1pm

Poster From

CURES-NOT-WARS

Dana Beal, Founder

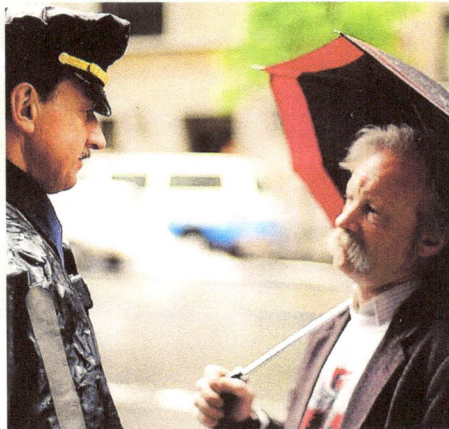

Hemp Hero Dana Beal speaking truth to government with guns at his annual Million Marijuana March in in New York City , 1998

WWW.CURES-NOT-WARS.ORG

"BOOK OF THE WORLD"

1853

From which the hemp
botanical drawing below
was taken

HEMP BOTANICAL DRAWING

HAND PAINTED

The female plant
is on the left,
male on the right,
hemp seeds below.
Hemp was legal
for all uses in 1853.

HEMP PLANTS

"HEMP IS GROWN IN AMERICA CHIEFLY FOR ITS COARSE FIBER, USED IN MAKING ROPES, CORDAGE, AND WARP FOR CARPET."

Hemp photograph from:

Luther Burbank: His Methods and Discoveries, Practical Applications - 1914, page 117

"ITS SEED IS LITTLE UTILIZED, ALTHOUGH IT MAKES A VALUABLE OIL."

43

HEMP AS AN EFFECTIVE ENERGY CROP, AND LOTS MORE, I.E. FOOD, PAPER, TEXTILES, PLASTICS, AND MEDICINE.

"At present there are comparatively few plants and a limited number of trees that will yield cellulose economically for quantity production. The plants are those containing vegetable fibers in the form of cotton, flax, hemp, jute, sugar cane, straw, espartro, and corn stalks. Of the trees there are spruce, balsam, fir, jack pine, hemlock, southern pine, poplar, and cottonwood." –Dard Hunter, PAPERMAKING, 1947, Borzoi Books.

HEMP for INDUSTRY

Willie Nelson's Biodiesel
Home of
Farm Fresh Biodiesel

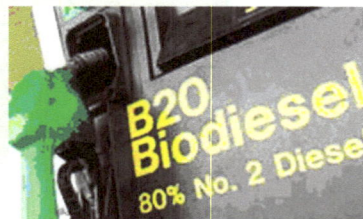

Put a B20 biodiesel blend in your tank and hit the road again with a clean burning, renewable fuel that is grown right here in America.

The Hemp Industries Association

A GROUP OF BUSINESS PEOPLE INTERESTED IN RESTORING HEMP TO THE U.S. "FREE MARKET"

"The Hemp Industries Association (HIA) is a non-profit trade group representing hemp companies, researchers and supporters. We are at the forefront of the drive for fair and equal treatment of industrial hemp. Since 1992, the HIA has been dedicated to education, industry development, and the accelerated expansion of hemp world market supply and demand.

The hemp industry has positioned itself over the past decade to once again become a major global economic force in the 21st century. Hemp is one of our planet's most important natural resources, and we advocate using it to its full potential.

If you are currently involved in the hemp industry, thinking of starting a hemp business, or support hemp commerce, please consider becoming a member."

There is no free market without hemp!

Our neighbor Canada

has now grown hemp for several years,

and England for almost ten years.

MEMBER
HIA
Hemp Industries Association

WWW.THEHIA.ORG

LEAVE THE DEPRESSION BEHIND YOU

BUY A FARM

Department of Agriculture

BUY A FARM

More than ever we need farmers to experiment with energy production, paper production, and resource development with hemp.

ORIGINAL ART BY:

MOLLY ENTNER COHEN

1935

5/23/1911 – 1/16/1986

LEAVE THE DEPRESSION BEHIND YOU

BUY A HEMP FARM

Department of Agriculture

BUY A HEMP FARM

By Molly Entner Cohen and Sherwood Akuna

Restoring the American family farm empowers small business and individuals to grow, process, and distribute the new energy, fiber, paper, food, and building products from hemp. Hemp can help heal the economy with its 50,000 plus products, fixing the damage done by NAFTA. A clean energy hemped economy also helps heal our planet.

USE HEMP TO LEAVE GLOBAL WARMING BEHIND US!

RASTA ARTIST SHERWOOD AKUNA UPDATED THIS CLASSIC WORK OF AMERICAN ART IN 2007

Farmers

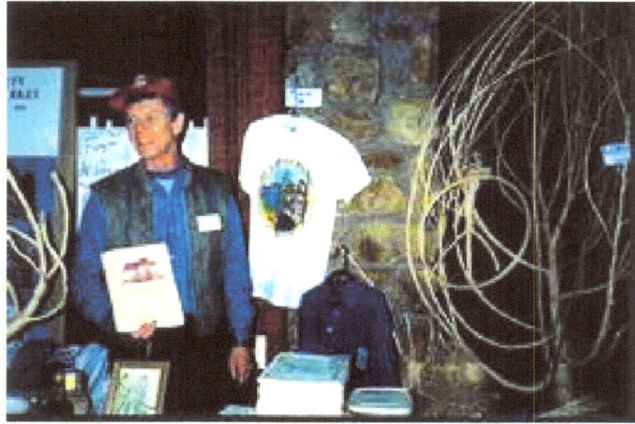

The Curator at a California Certified Organic Farmers meeting
with the beginnings of the Museum in 1991
flanked by hemp plants and a hemp peace symbol.

He was a member of CCOF for 13 years.

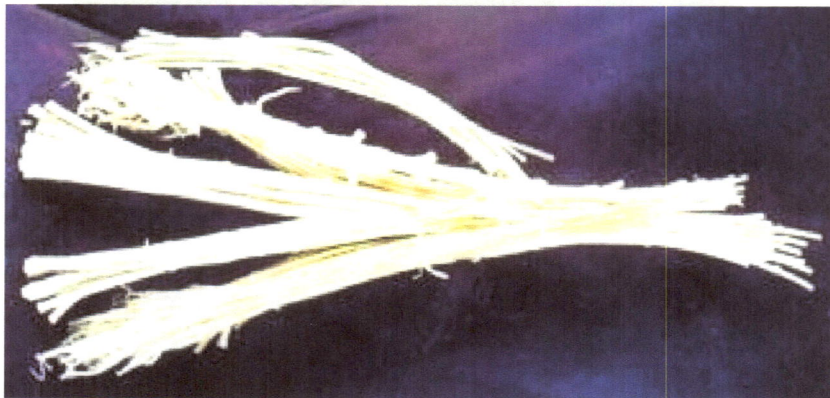

Hemp Museum dried hemp stalks, wild and domestic.

Bundling and tying hemp to be shocked.

Note the shocks in background.

This 'new' machine was designed to separate the fiber from the wood and do for hemp what the cotton gin did for cotton. One of the challenges of the global warming crisis is the development of new farm, biomass processing, and energy producing equipment. Other hemp growing countries have a head start over the U.S. in this development.

George Washington Carver

Carver was the greatest agricultural researcher
in American history.
He showed you could make almost anything
from farm products and agricultural waste.
Carver has been an inspiration to the USA Hemp Museum
to explore and research the hemp plant.

Carver developed hundreds of products out of the peanut.
Henry Ford used his ideas in building his early Ford cars.

GIVEN LOTS OF ROOM, HEMP PLANT BRANCHES
PROFUSELY FOR SEED OR MEDICINE.

HOW BIG DOES HEMP GROW IN ONE SEASON?
THIS MEDICAL - SEED PLANT
HAS A THREE INCH DIAMETER STALK.

IS IT ANY WONDER IT SKYROCKETS TO 15 FEET
IN A SHORT SEASON OF GROWTH
WHEN PLANTED CLOSE TOGETHER?

**TAP ROOTS OF WILD HEMP GROWN IN NEBRASKA
GREAT TO HOLD THE TOPSOIL
AND ADD HUMUS TO THE SOIL**

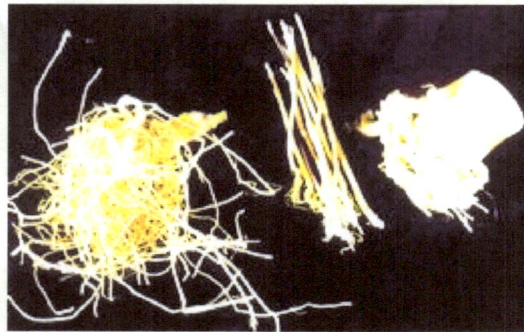

**SPREADING ROOTS OF A MEDICINAL –
SEED PLANT WITH LOTS OF ROOM BETWEEN PLANTS**

**CALENDAR COVER FOR THE KENTUCKY HEMP MUSEUM
SHOWING HISTORICAL PHOTO OF
SHOCKED BUNDLES OF HEMP (SHOCKS)
DRYING IN THE FIELD. 1997 CALENDAR**

**INTERNATIONAL
HEMP SEED STAMP
FROM CANADA**

CANNABIS, INTERNATIONAL

The Seed Supply Of The Nation - HEMP

Publication date: 1918 Source: 1917 Yearbook of the United States Department of Agriculture Author: R.A. Oakley, Agronomist in Charge of Seed Distribution, Bureau of Plant Industry Pages: 526-527

"Although we have still only a small acreage devoted to hemp in the United States, the acreage has doubled each year for the last three years. The area planted in 1917 was estimated at 42,000 acres. Kentucky supplies practically all of the hemp seed sown in this country. It is grown in seed plats along the Kentucky River. China and Japan furnish us large quantities of hemp seed for poultry feed, but it is practically valueless for seeding purposes. This seed can not be distinguished from our own domestic seed, and since it is much cheaper, fraud is often perpetrated on the unsuspecting farmer. The sale of Kentucky-grown hemp seed is controlled by such a small number of dealers that a tendency frequently develops toward the charging of exorbitant prices. Hemp must be specially planted for seed production, and in view of the increasing importance of the crop, seed production should be strongly encouraged. Chile offers possibilities in this connection, but for the present our efforts should be exerted at home. Our planting requirements, based on the acreage of 1917, are about 2,100,000 pounds of seed."

Hempseed Oil Products

HEMPSEEDS FROM CHINA

**Hemp seed is 30% oil by weight,
has a complete protein of very digestible quality.**

Besides its use in biodiesel, the seed oil of hemp has many other uses.

During the Congressional hearings on the 1937 Marijuana Tax Act, Ralph Loziers of the National Oil Seed Institute, representing paint manufacturers and high quality machine lubrication processors, showed up to disagree with the Act. He testified: "In the past 3 years there have been 193,000,000 pounds of hemp seed imported into this country, or an average of 64,000,000 pounds a year..." What is the oil used for, he was asked. "It is a drying oil, and its use is comparable to that of linseed oil or a perilla oil. It has a high iodine principle or strength. It is a rapidly drying oil to use in paints. It is also used in soap and in linoleum." (p. 61)

The hempseeds are pressed for oil, and contain about 30% oil by weight.

**IMPORTED
CANADIAN
HEMP OIL**

THIS ADVERTISING CAMPAIGN FOR ALTERNA COMPANY
LANDED THEM IN HOT WATER
WITH THE PLANT HATERS OF AMERICA.

WE THINK IT MAKES NO SENSE
THAT THIS HEMP FOR SHAMPOO,
THE FOOD I EAT, THE CLOTHES I WEAR,
CANNOT BE GROWN IN
THE UNITED STATES BY FARMERS HERE.

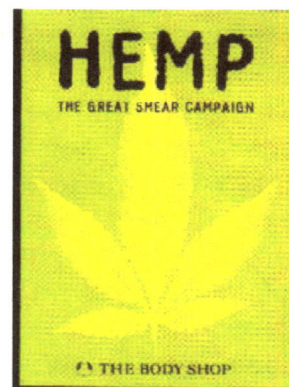

THE BODY SHOP HAD
THEIR OWN CAMPAIGN
AND WAS NOT SLOWED
IN THEIR GREAT
SMEAR CAMPAIGN.

WE WISH WE COULD USE SOME OF THIS OIL
TO GREASE THE WHEELS OF JUSTICE
TO LEGALIZE THE PLANT FOR AMERICA'S FARMERS.

WWW.HEMPMUSEUM.ORG

WWW.STARBULLETIN.COM

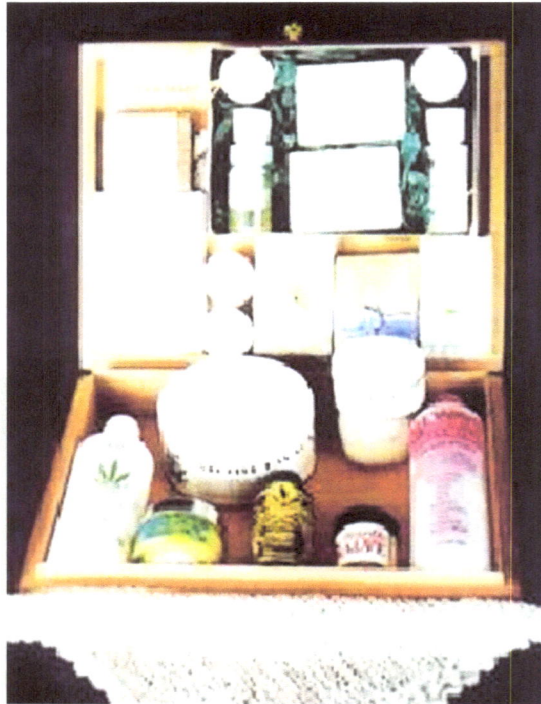

**SOME OF THE HEMPSEED OIL PRODUCTS
IN THE MUSEUM COLLECTION,
A WORLD MIX OF IMPORTED PRODUCTS.**

**LIP BALM, SOAPS, SALVE, SHAMPOOS,
SHOE CREAM, MASSAGE OIL, CRAYONS**

HEMP OIL LIP BALM

DR. BRONNER'S ALL-ONE EUCALYPTUS PURE-CASTILE HEMP SOAP

SEED OIL

WWW.GLOBALHEMPSTORE.COM

WWW.MANITOBAHARVEST.COM

WWW.HEMPOILCANADA.COM

WWW.NUTIVA.COM

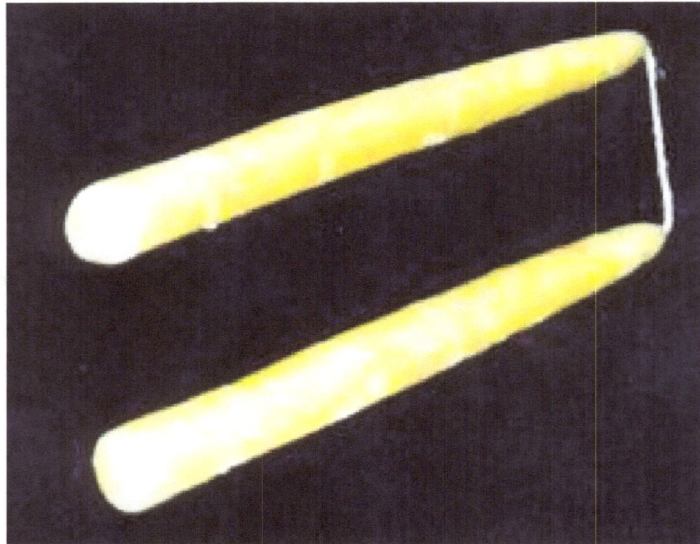

HEMPSEED OIL CANDLES AND HEMP WICKS

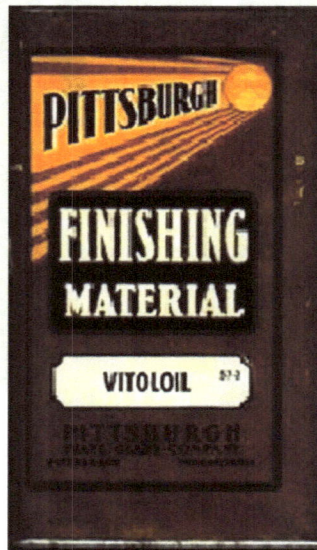

SOME PRODUCTS ARE HARD
TO FIGURE OUTTHE INGREDIENTS

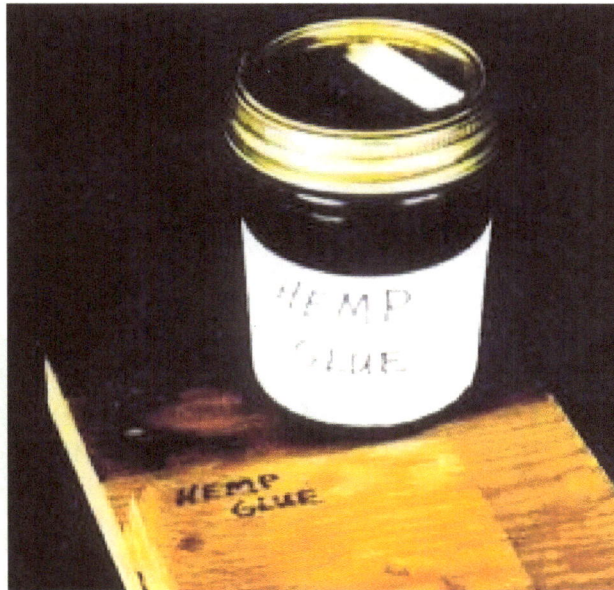

OTHER HEMP FEEDSTOCKS

The Curator made this lignin glue out of the liquid from the boiling pulp, which was boiled down until it was a thick boiling brown mass. The glue was water soluble, and seemed to absorb moisture from the air. When dry no one could pull the two boards apart with their hands.

For more information visit the USA Hemp Museum's Chemical Feed Stocks, Varnish & Paint, and Plastics Rooms at www.hempmuseum.org.

STILL GOOD

100 year old
can of
Hemp Varnish

6. THE WATER FACTOR

Plants, including forests and any plants such as hemp for biomass will require water to grow and thrive. It is likely then that water will become the new fossil fuel replacement – liquid gold that makes humanity survive and prosper.

The question of water usage in the State of California has a long and varied history. California has an incredible system of reservoirs, underground aquifers, canals and other water projects that capture about half of the annual runoff of some 70 million acre-feet of water. Agriculture takes 80 percent of captured runoff, leaving domestic, industrial, and environmental needs to vie for the remainder (see WATER: The Power, Promise, and Turmoil of North America's Fresh Water. Page 38, California: Desert in Disguise. (1993) National Geographic Special Edition.).

The water special above points out that we grow 700 square miles of rice in California irrigated with four to seven feet of water each year. Fifteen years ago, Larry Landis and a group called the Valley Keepers worked with the rice farmers of California's central valley who wanted to plant hemp in rotation with rice to control weeds. The weeds were controlled with herbicides at the time. The hemp was to be floated down river to the nearest paper mill and made into paper. Of course this was shot down by the Federal government.

These kinds of ideas must be given priority by the government. The time for drug war hysteria is over.

It is survival we are talking about and we must do strange and change now! Federal, state, and local governments must get this message. Do what needs to be done or let someone else do it. Free the hemp plant for every person in the world. Maybe we can save the world.

Another article in GREENPEACE (July-August 1989, by Marc Reisner author of Cadillac Desert) titled The Emerald Desert suggests we could steal (or borrow for a few years until global warming subsides) water from the cows. Wait until you hear this. Sprayed on grass (pasture) for cows raised in California is enough water for 20 million people. If we steal (or borrow for a few years until global warming subsides) the alfalfa crop from the cows, that is enough water for 20 million more people. In 1986, the gross state product for California was $575 billion; irrigated pasture's contribution was $94 million, one five-thousandth of the California economy and consumed one-seventh of the state water. Much of irrigated pasture's million acres could be planted with hemp.

62

And the entire cotton crop could be given over to hemp in this emergency survival situation where this can and should be done. Hemp produces twice as much fiber as cotton per acre, and the wood of hemp which is six times the weight of the hemp fiber can be used to make fuel, plastics, paper, etc. http://www.hempevolution.org/ecology/ecology.htm

"The primary ingredients for acid rain are sulfur dioxide and nitrogen oxides. We release these compounds into our atmosphere when we burn fuel. They mingle with water and oxygen particles in our atmosphere and create compounds. The compounds have an acidic pH level, and when they eventually fall this affects the earth in a variety of manners.

Last year we released 20 million tons of sulfur dioxide into the environment. When acid rain changes the pH of a lake or stream, the plants and animals can be harmed. Small food species like the mayfly cannot handle the change and will die out. Larger species that consume bugs like the mayfly (frogs, in this case) will also be affected. The whole ecosystem is in jeopardy. Animals like the clam cannot handle lower than pH 6. Meanwhile our lakes and streams are gradually getting more acidic. Little Echo Pond in New York has a pH of 4.2. If we continue this pace in the coming years, more of our precious resources will die out. There are already mounting levels of sulfur in our streams, lakes, and forests. Some lakes have no fish left at all. [xi]

When acid rain falls onto a forest floor the soil pH lowers. The whole ecosystem grows more slowly. While acid rain does not seem to affect trees directly, it can heavily damage roots and poison them. The sulfur dioxides can prevent vital nutrients from absorption. Acid rain releases aluminum and other toxic substances into the soil. Once the trees are weakened, they are more vulnerable to disease or insects, and even cold weather."

BURNING HEMP AND OTHER BIOMASS DOES NOT CAUSE ACID RAIN

7. THE POPULATION FACTOR

"It must be remembered, however, that in the long run, control of all greenhouse gases, including agriculture-related ones, will ultimately depend on success in controlling population growth." (Anne Ehrlich, Agricultural Contributions to Global Warming, p.400 of Global Warming by Greenpeace, 1990.)

"In the year 2050, there will be 120 million more Americans and we will be running out of open space...American farmland is disappearing at a rate of two acres every minute...Each day, America loses more than 6,000 acres of rural land to subdivisions, highways, and industrial malls." (2006, The Trust for Public Land)

**HEMP IS KEY TO SURVIVAL
OF OUR GLOBAL WARMING AND OTHER
ENVIRONMENTALLY INDUCED CRISIS SITUATIONS**

RE-HEMP

THE

PLANET

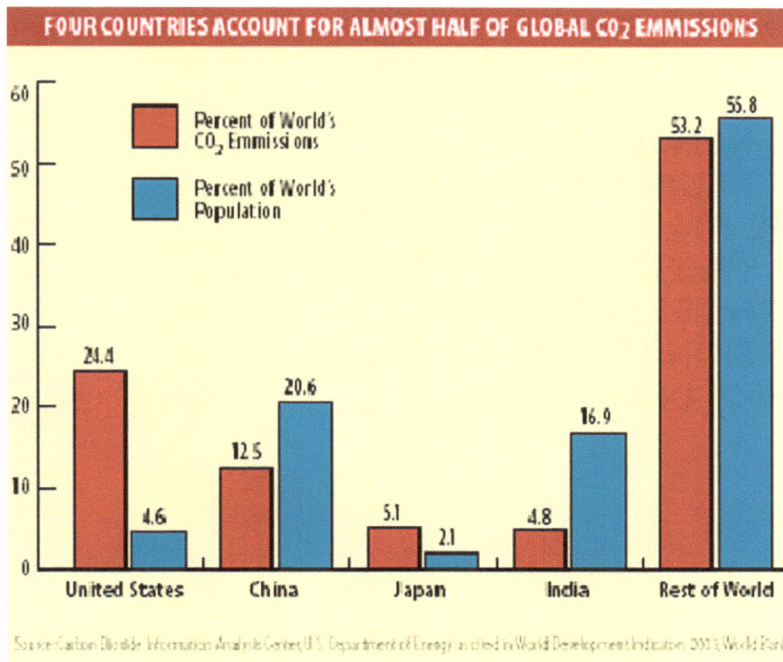

FOUR COUNTRIES ACCOUNT FOR ALMOST HALF OF GLOBAL CO_2 EMMISSIONS

Source: Carbon Dioxide Information Analysis Center, U.S. Department of Energy, as cited in World Development Indicators 2003, World Bank.

Two billion people in China and India are poised to copy our synthetic dead-end fossil fuel style that will lead to disaster on the global warming front. The United States can regain world respect by developing new ways of conservation, alternative energy products and ideas that are exportable, sustainable and renewable.

POPULATION CLOCKS

U.S. 300,972,251
WORLD 6,569,873,671
04:15 GMT (EST+5) JAN 15, 2007

U.S. 302,967,772
WORLD 6,620,570,807
16:12 GMT (EST+5) SEP 25, 2007

U.S. 304,160,373
WORLD 6,669,716,877
14:40 GMT (EST+5) MAY 24, 2008

WWW.CENSUS.GOV/MAIN/WWW/POPCLOCK.HTML

AUTOS, TRAINS & PLANES NEED: NEW, CLEAN BURNING FUELS. HIGH MILEAGE STANDARDS. LOWER SPEED LIMITS WITH TEETH. CONVERSION TO HEMP BIOFUEL

FACT: If America's autos were a separate country, they would be the world's fifth largest global warming polluter.

SIR. RICHARD BRANSON

"What sets climate change apart from these other crises is that most people can't see the problem -- CO2 gases are invisible.

If you could see them and they were colored red, 50 years ago it would have looked like a small brush fire smoldering around the world, and today it would look like a wildfire raging across the globe.

We desperately need leaders who can help bring visibility and forge solutions to this imperceptible menace before it's too late."

www.grist.org/news/maindish/2006/12/07/little

8. THERE WAS ANOTHER EMERGENCY WORLD WAR II

BASED ON THE FILM PRODUCED BY

THE U.S. DEPT. OF AGRICULTURE. 1942

COMIC BOOK FORMAT FOR FILM TEXT, 1990'S.

There was a similar emergency during World War II where hemp was brought back from Marijuana Prohibition to replace previously imported fiber for rope and military canvas. The crisis was so great that hemp farmers and their sons were exempt from military service. The Japanese military had taken over the Philippine Islands and cut off our supply of imported Manilla fiber.

Global warming is the kind of emergency that hemp can respond to better than other crops because it is biomass champion.

Hemp For Victory -1942

[Text of 14 minute U.S. Dept. Of Agriculture Film, 1942]

Long ago when these ancient Grecian temples were new, hemp was already old in the service of mankind. For thousands of years, even then, this plant had been grown for cordage and cloth in China and elsewhere in the Far East. For centuries prior to about 1850 all the ships that sailed the western seas were rigged with hempen rope and sails. For the sailor, no less than the hangman, hemp was indispensable. A 44-gun frigate like our cherished Old Ironsides took over 60 tons of hemp for rigging, including an anchor cable 25 inches in circumference. The Conestoga wagons and prairie schooners of pioneer days were covered with hemp canvas. Indeed the very word canvas comes from the Arabic word for hemp. In those days hemp was an important crop in Kentucky and Missouri. Then came cheaper imported fibers for cordage, like jute, sisal, and Manila hemp, and the culture of hemp in America declined.

But now with Philippine and East Indian sources of Hemp in the hands of the Japanese, and shipment of jute from India curtailed, American hemp must meet the needs of our Army and Navy as well as of our industry. In 1942, patriotic farmers at the government's request planted 36,000 acres of seed hemp, an increase of several thousand percent. The goal for 1943 is 50,000 acres of seed hemp. In Kentucky much of the seed hemp acreage is on river bottom land such as this. Some of these fields are inaccessible except by boat. Thus plans are afoot for a great expansion of a hemp industry as a part of the war program.

This film is designed to tell farmers how to handle this ancient crop now little known outside Kentucky and Wisconsin. This is hemp seed. Be careful how you use it. For to grow hemp legally, you must have a federal registration and tax stamp. This is provided for in your contract. Ask your county agent about it. Don't forget.

Hemp demands a rich, well-drained soil such as is found here in the Blue Grass region of Kentucky or in central Wisconsin. It must be loose and rich in organic matter. Poor soils won't do. Soil that will grow good corn will usually grow hemp. Hemp is not hard on the soil. In Kentucky it has been grown for several years on the same ground, though this practice is not recommended. A dense and shady crop, hemp tends to choke out weeds. Here's a Canada thistle that couldn't stand the competition, dead as a dodo. Thus hemp leaves the ground in good condition for the following crop. For fiber, hemp should be sewn closely, the closer the rows, the better. These rows are spaced about four inches. This hemp has been broadcast. Either way it should be sewn thick enough to grow a slender stalk. Here's an ideal stand: the right height to be harvested easily, thick enough to grow slender stalks that are easy to cut and process. Stalks like these here on the left yield the most fiber and the best. Those on the right are too coarse and woody.

For seed, hemp is planted in hills like corn. Sometimes by hand. Hemp is a dioecious plant. The female flower is inconspicuous. But the male flower is easily spotted. In seed production after the pollen has been shed, these male plants are cut out. These are the seeds on a female plant. Hemp for fiber is ready to harvest when the pollen is shedding and the leaves are falling. In Kentucky, hemp harvest comes in August. Here the old standby has been the self-rake reaper, which has been used for a generation or more. Hemp grows so luxuriantly in Kentucky that harvesting is sometimes difficult, which may account for the popularity of the self-rake with its lateral stroke. A modified rice binder has been used to some extent. This machine works well on average hemp. Recently, the improved hemp harvester, used for many years in Wisconsin, has been introduced in Kentucky. This machine spreads the hemp in a continuous swath. It is a far cry from this fast and efficient modern harvester, that doesn't stall in the heaviest hemp.

69

In Kentucky, hand cutting is practiced in opening fields for the machine. In Kentucky, hemp is shucked as soon as safe, after cutting, to be spread out for retting later in the fall. In Wisconsin, hemp is harvested in September. Here he hemp harvester with automatic spreader is standard equipment. Note how smoothly the rotating apron lays the swaths preparatory to retting. Here it is a common and essential practice to leave headlands around hemp fields. These strips may be planted with other crops, preferably small grain. Thus the harvester has room to make its first round without preparatory hand cutting. The other machine is running over corn stubble.

When the cutter bar is much shorter than the hemp is tall, overlapping occurs. Not so good for retting. The standard cut is eight to nine feet. The length of time hemp is left on the ground to ret depends on the weather. The swaths must be turned to get a uniform ret. When the woody core breaks away readily like this, the hemp is about ready to pick up and bind into bundles. Well retted hemp is light to dark grey. The fiber tends to pull away from the stalks. The presence of stalks in the bough-string stage indicates that retting is well underway. When hemp is short or tangled or when the ground is too wet for machines, it's bound by hand. A wooden bucket is used. Twine is used for the tying, but the hemp itself makes a good band. When conditions are favorable, the pickup binder is commonly used. The swaths should lie smooth and even with the stalks parallel. The picker won't work well in tangled hemp. After binding, hemp is shucked as soon as possible to stop further retting. In 1942, 14,000 acres of fiber hemp were harvested in the United States. The goal for the old standby cordage fiber [about 360,000 acres], is staging a strong comeback.

This is Kentucky hemp going into the dryer over mill at Versailles. In the old days braking was done by hand, one of the hardest jobs known to man. Now the power braker makes quick work of it. Spinning American hemp into rope yarn or twine in the old Kentucky river mill at Frankfort, Kentucky. Another pioneer plant that has been making cordage for more than a century. All such plants will presently be turning out products spun from American-grown hemp; twine of various kinds for tying and upholster's work; rope for marine rigging and towing; for hay forks, derricks, and heavy duty tackle; light duty fire hose; thread for shoes for millions of American soldiers; and parachute webbing for our paratroopers. As for the United States Navy, every battleship requires 34,000 feet of rope. Here in the Boston Navy Yard, where men are now working night and day making cordage for the fleet. In the old days rope yarn was spun by hand. The rope yarn feeds through holes in an iron plate. This is Manila hemp from the Navy's rapidly dwindling reserves. When it is gone, American hemp will go on duty again: hemp for mooring ships; hemp for tow lines; hemp for tackle and gear; hemp for countless naval uses both on ship and shore. Just as in the days when Old Ironsides sailed the seas victorious with her hempen shrouds and hempen sails. Hemp For Victory.

THE END

HEMP HERO

PETER

MCWILLIAMS

1950-2000

From Ain't Nobody's Business If You Do

Part V: WHAT TO DO? - HEMP FOR VICTORY

"One of the most beneficial aspects of using hemp (or other plants) for fuel is that, as plants grow, the plants take carbon dioxide out of the atmosphere and replace it with oxygen. This helps solve one of our primary environmental problems: too much carbon dioxide. When a portion of the hemp plant is burned for fuel, it has already "earned" the oxygen it uses by having placed that oxygen in the atmosphere while it grew. Fossil fuels (oil, gas, or coal), on the other hand, come from plant and animal sources that died millions of years ago—whatever carbon dioxide they took or oxygen they left happened millions of years ago. Burning fossil fuels, then, only adds to carbon dioxide and reduces the amount of oxygen in the atmosphere."

WWW.MCWILLIAMS.COM

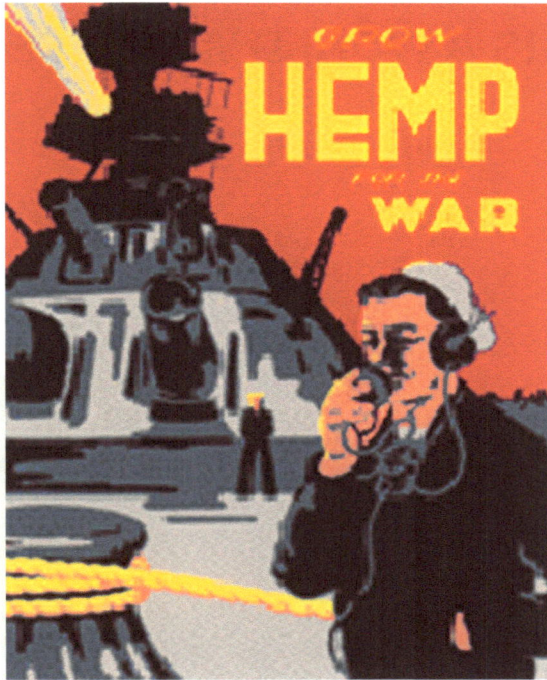

WHAT IS HEMP FOR VICTORY: A GLOBAL WARMING SOLUTION?

A PLAN TO USE HEMP AS A CLEAN BIOFUEL
TO REPLACE TOXIC FUELS.

A PLAN TO GROW HEMP TO SCRUB THE AIR OF
THE EXCESS GREENHOUSE GAS CO_2
ON AT LEAST 10% OF UNUSED FEDERAL LAND.

A PLAN TO LEGALIZE AND TAX HEMP TO CREATE A FUND TO
DEAL WITH GLOBAL WARMING AND ITS EFFECTS.

A PLAN TO DEVELOP HEMP'S
50,000 PLUS VIABLE USES

HEMP IS A WORLD WAR II VETERAN

71 Hemp Mills Planned

THE COMMODITY CREDIT Corp. last month made known a processing and production program which will entail the building of 71 hemp mills in this country in 1943, with an expected production of some 150,000 tons of fiber for essential military and civilian cordage.

The plants will be operated by a newly-created Commodity Credit Corp. division headed by Samuel H. McCrory, formerly of the Bureau of Agricultural Chemistry and Engineering.

Location of the plants will be determined by the signing of contracts by farmers for the production of hemp fiber in 1943. Each plant will be situated so as to serve approximately 4,000 acres of hemp, and most are expected to be located in Kentucky, Indiana, Illinois, Wisconsin, Iowa, and Minnesota.

71 HEMP MILLS PLANNED

THE COMMODITY CREDIT CORP. last month made known a processing and production program which will entail the building of 71 hemp mills in this country in 1943, with an expected production of some 150,000 tons of fiber for essential military and civilian cordage.

The plants will be operated by a newly-created Commodity Credit Corp. division headed by Samuel H. McCrory, formerly of the bureau of Agricultural Chemistry and Engineering.

Location of the plants will be determined by the signing of contracts by farmers for the production of hemp fiber in 1943. Each plant will be situated so as to serve approximately 4,000 acres of hemp, and most are expected to be located in Kentucky, Indiana, Illinois, Wisconsin, Iowa, and Minnesota.

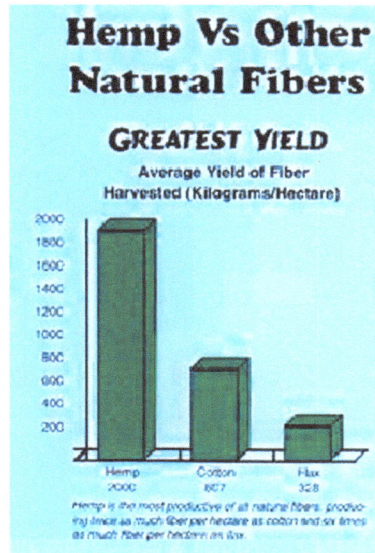

Hemp Vs Other Natural Fibers

GREATEST YIELD

Average Yield of Fiber
Harvested (Kilograms/Hectare)

Hemp 2000	Cotton 807	Flax 328

Hemp is the most productive of all natural fibers, producing twice as much fiber per hectare as cotton and six times as much fiber per hectare as flax.

"Hemp is the most productive of all natural fibers, producing twice as much fiber per hectare as cotton and six times as much fiber per hectare as flax."

WWW.AURORASILK.COM

9. A NATURAL ENERGY POLICY

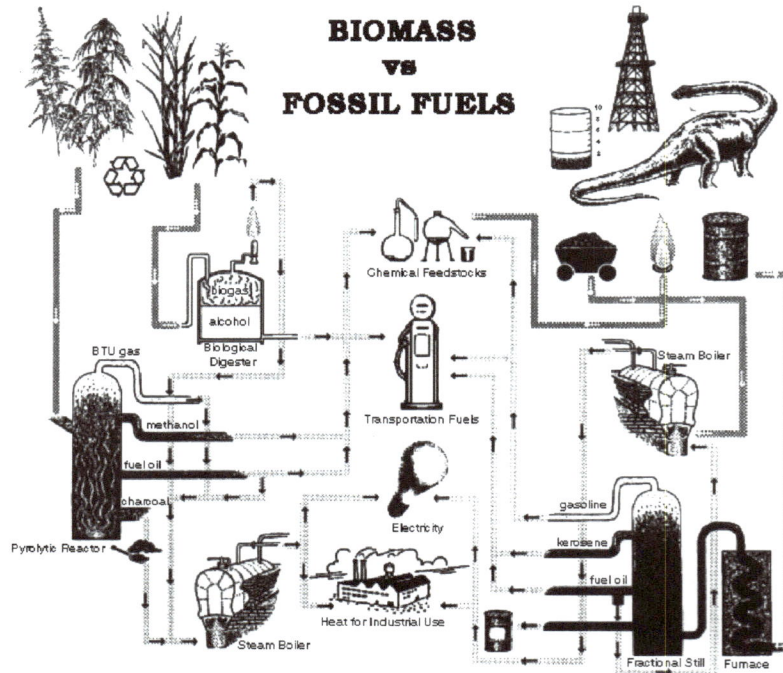

**BIOMASS
vs
FOSSIL FUELS**

(Above graphic from www.globalhemp.com)

Originally written by the Curator in 1992

Something has happened on the alternative energy front that is so revolutionary that all people connected with or interested in improving the quality of life on our planet should be aware of it. A solar collector has been re-discovered (you probably think this is a joke). Once declared useless by our government in 1937, this collector is so powerful it could *replace every type of fossil fuel energy product* (oil, coal, and natural gas).

This solar collector is a green plant, one of the most advanced in the plant kingdom. It uses the evenly distributed light of the sun to grow biomass (biologically produced matter). This plant is the *earth's number one biomass resource* or fastest growing annual plant for agriculture on a worldwide basis, producing up to 14 tons per acre. This is the *only biomass source* available that is capable of producing all the energy needs of the U.S. and the world.

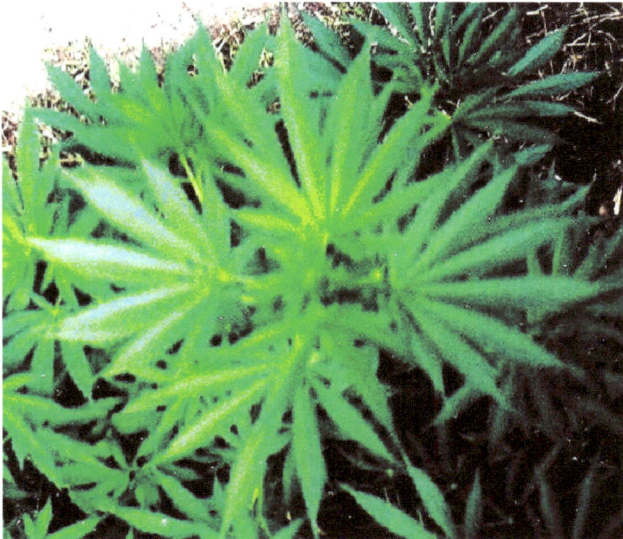

We need to gain a vision of where we have to go to heal our home, stop the poisons, stop the wars, learn the natural ways, and learn to love our common home and our sisters and brothers.

We need local, family owned, energy farms to lift us out of the death-like grip of big oil; and give promise to future generations of a renewable, sustainable energy source.

Fuel is not synonymous with petroleum, *let's get over that.* New annually renewable biomass energy systems will create millions of new jobs.

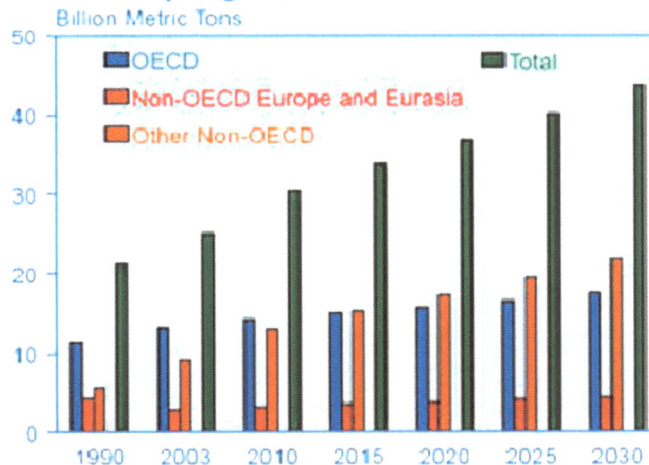

Figure 65. World Carbon Dioxide Emissions by Region, 1990-2030

Sources **1990 and 2003**: Energy Information Administration (EIA), *International Energy Annual 2003* (May-July 2005), web site www.eia.doe.gov/iea/. **2010-2030**: EIA, System for the Analysis of Global Energy Markets (2006)

"DIE FOR PETROLEUM OR LIVE WITH HEMP

(CHOOSE ONE)"

In case you doubt the power of this miracle plant, consider what else it can do: Replace all wood pulp paper products with a far superior, dioxin-free paper. Provide the strongest textiles, ropes, fabrics, and fibers for clothing (it is softer than cotton); Provide time tested and safer medicines for a hundred or more different medical conditions; Provide high protein food stuffs (soybeans alone have a bit more protein) and high quality vegetable oil (with heart helping Omega 3 fatty acids like fish oil); Provide raw materials for plastics and building materials like composition board; Provide raw materials for 50,000 commercial uses that are economically viable and market competitive; And, oh yes, provide a safe, sane, non-violent recreational drug. For an energy policy we can live with and flourish with for years to come, think hemp. Think *Cannabis sativa*, the plant. Let's allow competition in the best free market sense. Put it out there, let it fly and be free. Free at last.

Back to energy. Why worry about energy? Let me get your attention: According to The Emperor Wears No Clothes, eighty per cent (80%) of the total dollar expense of living for each human being is energy cost. That means that 33 hours of each 40 hour work week goes to pay for energy costs in goods and services, whichever way (manufacturing, transportation, heating, cooking, lighting, etc) you purchase.

78

Here is a quote from "The Emperor..."

"Our current fossil energy sources also supply about 80% of the solid and airborne pollution which is slowly poisoning the planet. (See U.S. EPA report 1983-89 on coming world catastrophe from carbon dioxide imbalance caused by burning fossil fuels - (oil, coal, and natural gas) [now called the greenhouse effect]. The cheapest substitute for these expensive and wasteful energy methods is not wind or solar panels, nuclear; geothermal, and the like, but using the evenly distributed light of the sun to grow plant biomass."

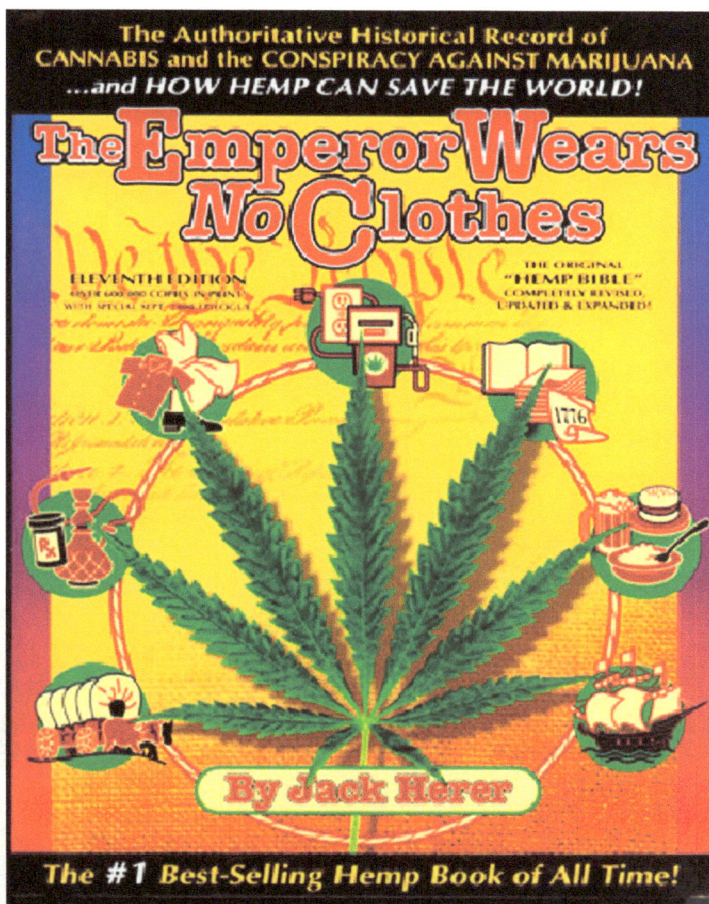

The Authoritative Historical Record of CANNABIS and the CONSPIRACY AGAINST MARIJUANA ...and HOW HEMP CAN SAVE THE WORLD!

The Emperor Wears No Clothes

ELEVENTH EDITION

THE ORIGINAL "HEMP BIBLE" COMPLETELY REVISED, UPDATED & EXPANDED!

By Jack Herer

The #1 Best-Selling Hemp Book of All Time!

A decade later another Bush President refuses to cooperate with global concerns on global warming.

The American farmer has been displaced by the synthetic fossil fuel people, and we have all paid the price. Who do we want to give our energy dollars to?

We need an exportable, ecologically sound lifestyle to sell to the world. We are a world at need, for food, clean water, shelter, and energy (clean, renewable, natural, almost universal energy from hemp). George Bush had us fight an oil war at a cost of American lives, Iraqi lives, and $61 billion, to save a lifestyle (synthetic dead-end fossil fuel style) that is not only not exportable but rapidly ruining our country as well. [Now another Bush is fighting another oil war. The cost has risen to hundreds of billions of dollars and at this date (Sept., 2006) 2,600 American lives have been lost in that war.]

"A co-generation system for converting walnut shells into energy was built by Diamond Walnut Growers to supply power for its Stockton

79

plant. The cooperative also markets energy to local utility companies." - *California Agriculture*, University of California, 1983.

The Cannabis hemp/marijuana movement is not an undercurrent in this country; it is an undertow. We are going to flood the American people with American hemp history and pride in our connection with this plant. I want you to understand the hard truths that marijuana prohibition has obscured. This plant (or any plant) should never have been made illegal. Our first flag was made of Cannabis hemp/marijuana. Our constitution was written on hemp paper. The facts are in Jack Herer's book: *The Emperor Wears No Clothes: Hemp and the Marijuana Conspiracy*, [and *Hemp: Lifeline to the Future*, by Chris Conrad]. These books are required reading for every American to learn the lost history of hemp, hemp/marijuana prohibition, and how hemp can save the world from energy madness, if we can act immediately to put hemp back into the free market - your market. Then farmers can plant our nation's fuel, fiber, paper, medicine, food, plastic and future. It is our choice. How long will we have to wait to establish a sane and survivable lifestyle, based on the natural cycles such as that of hemp?

This is the ecological truth: The sooner we act to end our synthetic society the less damage to earth. We Americans - 5% of the world population - in our drive for more "net worth" and "productivity" use 25% to 40% of the world's energy. As a country, we have been horribly deceived for the past 69 years (1937-2006). Long enough. Never again. The repression of information about hemp has cost the U.S. about 80% of our petroleum reserves. Add to that the 70% of our forests that did not have to be cut down for making paper. Add to that too many family farms gone. Add to that the 50,000 Americans and the 10,000 Canadians killed annually by acid rain from burning high-sulfur coal.

The world struggle for money is actually a struggle for energy, as it is through energy that we may produce food, shelter, transportation and entertainment. As we have seen with the Bush Administration, it is this struggle which often erupts into open war. Ultimately, whether from too much pollution, too many wars, too high a price, the world has no other rational environmental choice but to give up fossil fuels.

Because of the second prohibition that surrounds Cannabis hemp/marijuana, we are not told the truth even by our own U.S. Environmental Protection Agency scientists. Hemp is the home grown, annually renewable, CHEAPEST source of energy.

"Rather than as a crisis, the energy problem can be viewed as a challenge and opportunity." - 1983. *California Agriculture.*

Keep in mind the excellent properties of Cannabis hemp/marijuana in reading the following article. Hemp is clearly the ecological and economical choice. "*The Case for Methanol*," printed in *Scientific American*, November, 1989, was written by two U.S. Environmental Protection Agency scientists, Charles L. Gray, Jr., and Jeffery A. Alson. The authors maintain that a move to pure methanol fuel would reduce vehicular emissions of hydrocarbons and green house gases and could lessen U.S. dependency on foreign energy sources. Here are a few paragraphs from the article, obtainable from any library:

"The private automobile has shaped U.S. society to a degree unparalleled by any other product of the industrial age. By providing mobility and convenience particularly attuned to the American desire for personal freedom, the automobile has come to dominate not only the nation's transportation network but also its very culture. And the automotive industry has become a pillar of the economy, accounting for more than 10 percent of the gross national product and some 20 percent of all consumer expenditures. Yet the automobile ALSO THREATENS the quality of life, contaminating both urban air and the global atmosphere, where automobile emissions contribute to the green house effect. The automotive industry must overcome unprecedented technical, political and social challenges if these serious environmental problems are to be solved.

To achieve this goal, <u>we believe the nation must begin making a transition to a new automotive fuel.</u> Having studied a wide range of alternatives, we think that fuel should be methanol [wood alcohol]. A move to methanol could achieve emission reductions far beyond those that are feasible even with advanced emission controls on gasoline vehicles. Although the past 15 years have seen substantial reductions in noxious pollutants and greenhouse gases from individual vehicles, the number of vehicles has been steadily increasing. Consequently, more than 100 cities still have ambient levels of carbon monoxide, particulate matter and ozone (generated from photo-chemical reactions with hydrocarbons from vehicle exhaust) that exceed the levels established by the Environmental Protection Agency to protect public health. As the nation's fleet continues to grow in the next decade, air quality will worsen unless vehicles can be developed that are much clearer than those on the roads today.

Introducing methanol to the U.S. transportation infrastructure would require relatively modest changes for the automotive and energy industries. Our research has convinced us that this is the only practical means to achieve major reductions in vehicle emissions while maintaining the personal mobility that Americans have come to expect.

Although there will be costs in making such a transition, there will also be significant benefits not only for the environment but most likely for the nation's economic health as well. We have incontrovertible evidence from vehicle tests and computer simulations that vehicles operating on pure methanol would bring about dramatic decreases in urban levels of ozone and toxic substances. What is more, methanol can be produced with current technologies from a variety of abundant sources, including natural gas, coal, wood and even organic garbage [and the cheapest source, HEMP]. By beginning a transition to methanol, the nation could ultimately lessen its dependence on foreign sources of energy."

GRAPHIC TEXT
1992 FLEXIBLE FUEL METHANOL ECONOLINE VAN

**These specifications are supplied
by Ford Motor Company
and may be changed without notice.
Specific fuel flexible vehicle option
will be available in October
through dealer who is
awarded state contract...**

1992 FLEXIBLE FUEL METHANOL ECONOLINE VAN

These specifications are supplied by Ford Motor Company and may be changed without notice.

Specific fuel flexible vehicle option list will be available in October through dealer who is awarded state contract.

Please contact California Energy Commission's Transportation Technologies and Fuels Office for additional information at (916) 654-4602.

**Please contact
California Energy Commission's
Transportation Technologies and Fuels
Office for addressing information at
www.energy.ca.gov**

**Ford Van runs on any mix
of gasoline and alcohol.
California Energy Commission, 1992**

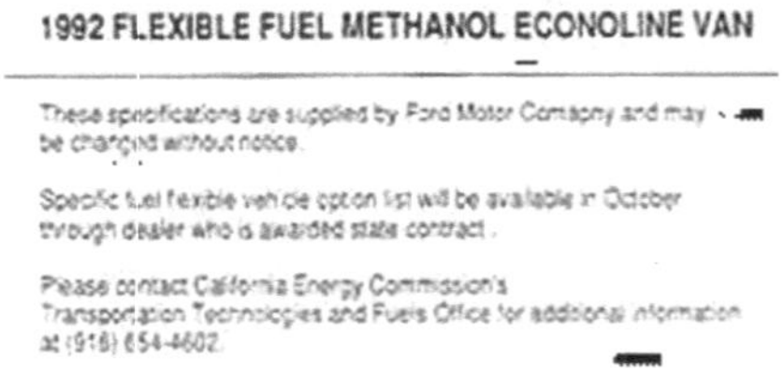

From *The Emperor...*, p.43: "The biomass conversion process can produce [ethanol], methanol, fuel oil, charcoal fuel, as well as the basic chemicals of industry: acetone, ethyl acetate, tar, pitch, and creosote. The Ford Motor Company successfully operated a biomass "cracking" plant in the 1930's at Iron Mountain, Michigan...Henry Ford even grew Cannabis hemp/marijuana on his estate after 1937, possibly to prove the cheapness of methanol production."

82

[On January 28, 1999, The Los Angeles Times printed a special section called "Highway 1" which featured innovations in technology about autos.]

"Fuel Cell Technology: Fuel cells are being called the best possible source of power for the electric car of the future. Car companies are spending billions of dollars on development. Fuel cells use a chemical reaction to produce electricity from hydrogen, which can be stored in tanks in the vehicle or distilled from gasoline, methane, and other hydrocarbon-based fuels.

Again from *The Emperor...*, look what happened when we had a national emergency in World War II, the most recent time

America asked its farmers to grow more Cannabis hemp/ marijuana:

"Our national energy needs are an undeniable national security priority. Look what Uncle Sam can do when pushed into action:

In 1942, Japan cut off our supplies of vital hemp and course fibers. Cannabis hemp/marijuana which had been outlawed as the "Assassin of Youth" just four years earlier was suddenly safe enough for our government to ask the kids in the Kentucky 4H Clubs to grow at least half an acre but preferably two acres of hemp each. (U. of KY Ag. Extension Leaflet 25, Mar., 1943.)

In 1942-43 farmers were made to attend showings of the USDA film "Hemp for Victory," [Text printed herein] sign that they had seen the film, and read a hemp cultivation booklet. Hemp harvesting was made available at low or no cost. Five dollar tax stamps were available and 360, 000 acres of cultivated hemp was the goal by 1943.

Farmers from 1942 through 1945 who agreed to grow hemp were waived from serving in the military, along with their sons; that's how vitally important hemp was to America during World War II."

I have said this before, and here is the proof of what we did in a crisis, the American farmer is two years away from a major hemp crop. We are in an environmental crisis of enormous proportions, and instead of asking the American farmer to assist again, George Bush led us to war [as his son is now doing]. This is just another one of those situations in which George Bush is damned if he knew about hemp and didn't employ it, and damned if he just didn't know.

Al Gore, Environmental Activist

**THE MAN WHO WON THE POPULAR VOTE
FOR PRESIDENT OF THE USA IN 2000.**

**"I'VE BEEN INCREDIBLY GRATIFIED
BY THE RESPONSE TO AN INCONVENIENT TRUTH.
I'M EXTREMELY PROUD OF ALL THE WORK
THE TEAM PUT INTO THE FILM
AND IT FEELS LIKE IT CAME AT A CRUCIAL TIME.**

**BUT NOW COMES THE HARD WORK.
WE HAVE TO TAKE THIS MESSAGE TO WASHINGTON.
AND WE CAN'T DO IT WITHOUT YOU."**

"Humanity is sitting on a ticking time bomb. If the vast majority of the world's scientists are right, we have just ten years to avert a major catastrophe that could send our entire planet into a tail-spin of epic destruction involving extreme weather, floods, droughts, epidemics and killer heat waves beyond anything we have ever experienced. "

"It is difficult to get a man to understand something when his salary depends on his not understanding it."

WWW.WECANSOLVEIT.ORG

By far the most terrifying film
you will ever see.
aninconvenienttruth
A GLOBAL WARNING

Speaker of the House Nancy Pelosi

2007 Website Posting
WWW.SPEAKER.HOUSE.GOV

Energy Independence

Energy independence is a national security issue, an economic issue, and an environmental issue. With gasoline prices at record levels, Americans are feeling the pain at the pump. They worry about the security of our nation and our growing dependence on foreign oil. Fortunately, the answer to this long-term challenge is right here at home.

American ingenuity can be put to work to achieve energy independence from Middle East oil in the next 10 years. America can develop emerging technologies to process homegrown alternative fuels. A sustained investment in research and development is crucial to creating cutting-edge technologies to develop these clean, sustainable energy alternatives and capitalize on America's vast renewable natural resources, including solar energy and wind power.

Democrats have a plan developed by our Rural Working Group to invest in the Midwest and other American farm communities. Our New Direction will send our energy dollars to the Midwest, not to the Middle East.

With solutions that are home-grown, our plan commits to America. Our comprehensive plan provides tax incentives to encourage increased biofuels production, increases the number of flex fuel vehicles on the road, and expands the ethanol pumps at gas stations. It would increase research and development to create jobs through cutting-edge technologies for biofuels, including new refining processes and new vehicle technologies.

The technological advances that will achieve energy independence also will help us address the most urgent environmental issue facing us today: global warming. For the sake of our future generations, America must provide strong leadership to reduce emissions that are responsible for global warming. Instead, we have walked away from international efforts to help reduce this growing danger to the planet.

We will develop ground-breaking technology and policies that harness the creativity and flexibility of the free market to reverse the dangerous warming trends.

Congressman Dennis Kucinich

Halt global warming !
Legalize hemp today!!

CANNABIS HEMP
IS THE SINGLE MOST IMPORTANT TOOL
TO REVITALIZING OUR ENVIRONMENT
AND PROMOTING NEW
GLOBAL CONCERN

WWW.KUCINICH.US

Many people know that Cannabis Hemp will be good for our environment. IT has been endorsed by various political & environmental groups. There are a variety of ways that Cannabis Hemp can promote environmental awareness and help restore our earth. Until now, however, Cannabis Hemp has been a low priority among environmental activists, especially due to its association with Marijuana. Here on this website are the summary uses of Cannabis Hemp for our environment. No one should doubt this vital information. There is no better solution to our ecological problems. The truth is as plain as day.

GLOBAL WARMING and CANNABIS HEMP

Global warming is not a myth. Many people see Global Warming as the most threatening force of destruction on our planet. Global Warming is linked to a number of other environmental problems affecting the earth. Millions of people would die as a result of global warming if no changes were made.[i]

The threat of global warming has already prompted large money and time investments on the part of environmental watchdog groups like the Sierra Club.

Briefly summarized, global warming is the increasing buildup of Carbon Dioxide in our atmosphere. Carbon Dioxide is released when fossil fuels, such as coal or petroleum, are burned for energy. In the last

100 years alone we have increased our CO2 levels by 30%[ii], with a noticeable effect on the environment. Our sea levels are rising, the global temperature is increasing, our glaciers are melting, and scientists predict further and more massive levels of destruction in the coming years. The dependence on petroleum products and fossil fuels could potentially ruin the earth, making it possibly uninhabitable in the coming centuries.

Power plants release carbon dioxide when they produce energy. In 1998 electric utilities released about 550 million tons into the atmosphere.[iii] Because the carbon dioxide comes from energy that has been stored for millions of years, this adds an unexpected burden to the environment. While plants help a little by taking CO2 out of the atmosphere, there is far too much for the plants to take it all in.

Burning Cannabis Hemp for energy would solve this problem. Cannabis Hemp is a plant, and gets its energy from the sun. This process, called photosynthesis, produces oxygen and takes carbon dioxide from the atmosphere. An increase of plant growth both domestically and abroad would lower the CO2 levels in our atmosphere, and promote a healthy environment. Growing Cannabis for other goods (like fibers) would further decrease our excess CO2 burden.

It is possible to produce all of our energy with Cannabis Hemp. The unique growing properties of the plant make it the ideal crop for our energy needs. One acre of Cannabis Hemp can produce 1000 gallons of methanol in a single growing season. Any CO2 released from burning Cannabis Hemp would be the same CO2 the plant had already taken from the environment, creating what is called a closed carbon cycle. A closed carbon cycle system of energy production would slow down the effects of global warming, and with well-implemented plant growth could possibly stop global warming entirely.[iv] No other plant on earth could meet the needs of global energy consumption, but Cannabis Hemp could.

Our automobiles account for much of the CO2 released into the ecosystem. Already electric cars are available to the public to promote reduced gasoline consumption. But electric cars are inefficient, and the support structure for this type of automobile is not in place. In 1998 transportation fuels accounted for almost 500 million tons of CO2 emissions.[v] Meanwhile, our gas prices are skyrocketing, taking money directly from our pockets.

Cannabis Hemp can produce a clean-burning, energy efficient form of gasoline, with less cost to the consumer. Already, ethanol is added to gasoline to increase octane levels and efficiency. Henry Ford of Ford motors believed that eventually all cars would run entirely on

88

ethanol. Cannabis Hemp can easily be fermented into alcohol in the form of ethanol. Small ethanol production stills from corn and other crops already exist in the United States. Making ethanol and gasoline from Cannabis Hemp would further reduce CO2 emissions and help regenerate our suffering planet. The National Renewable Energy Lab in Colorado, the Environmental Protection Agency, and the U.S. Department of Energy have all stated that to help the environment, we must produce biodiesel and bioethanol.

If ethanol production proves too daunting, and more immediate solutions are warranted, cannabis hemp can also be converted into fuel oils to produce gasoline directly. Cannabis gasoline would not contain other harmful emissions associated with automobiles (such as sulfur), but it would release carbon dioxide. Still, as mentioned earlier, this CO2 would be the same CO2 the cannabis plant had already absorbed, and so there would be no net increase. Either way, Cannabis Hemp fuels will be essential in the new century.

It is essential that Cannabis Hemp be used to produce energy. Our environment is precious and we have the potential to stop the destruction. Our global ecosystems cannot wait. Support Cannabis Hemp for renewable energy and help save our environment.

DEFORESTATION and CANNABIS HEMP

The timber industry has long been essential to produce jobs and manufacture products in our country. This has been, alas, at great expense to our environment. Logging destroys forests, hurts streams, kills animals and plants, wipes out species, and pollutes our environment, to name just a few of its problems. Still, the resource has been essential through modern day and so we keep cutting down trees. Now it is time to stop, before we lose more of our precious heritage.

Today we make 93% of our paper from trees, including cardboard, printing paper, newspaper, etc. We use almost 40% of our forests for timber.[vi] This fills our water with nitrates, and that has terrible effects on the ecosystem. One quarter of our forests are critically imperiled, meaning they are vulnerable or unique. Many of these critically imperiled forests are not currently protected and could be destroyed any time. This problem even reaches into our wallets. We gave the Forest Service two billion dollars in subsidies from 1992-1997. This means we are actually paying taxes to destroy our own trees and heritage.[vii]

Cannabis Hemp can replace any of the products made from timber. No more forests would be needlessly wasted. This would save precious resources and renew the ecosystems. More importantly, it would mean more beautiful heritage to grow for our children.

All the paper we make from trees could be made better by using Cannabis Hemp. We would make more paper per acre. Each Cannabis plant grown saves 12 trees. Cannabis Hemp uses about 1/7 the chemicals in paper manufacture. Right now we cut down about 500 million cubic meters of forest every year.[ix]

Cannabis could be used for particleboards of any size, as well as insulation, drywall, cabinets, and furniture. We could build a house from Cannabis Hemp materials without excess pollution, and without cutting down a single tree.

Logging our trees is without cause. There is no more need to waste our land when Cannabis Hemp could easily replace our timber. This destruction reaches into the homes and pockets of every taxpayer, and meanwhile our children cannot play in the streams and the logging industry gets another needless government refund. Now it is time to turn to the future. Hemp is our solution.

ACID RAIN and CANNABIS HEMP

Acid Rain comes from byproducts released in the fumes when we burn fossil fuels (petroleum, coal, etc.). Acid rain affects our environment in several ways, and can harm humans as well. Acid rain is also damaging to buildings and structures; it decays unprotected monuments and statues important to our cultural heritage.[x]

The primary ingredients for acid rain are sulfur dioxide and nitrogen oxides. We release these compounds into our atmosphere when we burn fuel. They mingle with water and oxygen particles in our atmosphere and create compounds. The compounds have an acidic pH level, and when they eventually fall this affects the earth in a variety of manners.

Last year we released 20 million tons of sulfur dioxide into the environment. When acid rain changes the pH of a lake or stream, the plants and animals can be harmed. Small food species like the mayfly cannot handle the change and will die out. Larger species that consume bugs like the mayfly (frogs, in this case) will also be affected. The whole ecosystem is in jeopardy. Animals like the clam cannot handle lower than pH 6. Meanwhile our lakes and streams are gradually getting more acidic. Little Echo Pond in New York has a pH of 4.2. If we continue this pace in the coming years, more of our precious resources will die out.

There are already mounting levels of sulfur in our streams, lakes, and forests. Some lakes have no fish left at all.[xi]

When acid rain falls onto a forest floor the soil pH lowers. The whole ecosystem grows more slowly. While acid rain does not seem to affect trees directly, it can heavily damage roots and poison them. The sulfur dioxides can prevent vital nutrients from absorption. Acid rain releases aluminum and other toxic substances into the soil. Once the trees are weakened, they are more vulnerable to disease or insects, and even cold weather.

Acid rain affects more than our natural environment. Our statues and monuments, our buildings and houses, all deteriorate over time, and acid rain speeds this process. Statues like the Parthenon lose their features and will never be replaced. Acid rain decays the marble on our steps and columns, and the metal on our buildings. Replacing damage from acid rain can cost billions of dollars.[xii]

There is no reason for us to poison our own planet. Clean energy from Cannabis Hemp would totally solve this problem. Cannabis Hemp does not contain high levels of nitrogen or sulfur. If we burned Cannabis Hemp for electricity and gasoline we would stop releasing sulfur compounds. Because Cannabis Hemp grows very rapidly and is easily harvestable, we could produce all the energy and gasoline we needed from Cannabis Hemp. When that happens, we can start to rebuild our decaying forests and structures without fear that they will be destroyed again.

There are a number of problems caused by needless fossil fuel use. Cannabis Hemp for energy is the necessity of the future and will soon be in place all over the world. When acid rain goes away the forests will become healthy again. Nature will begin to resume its balance.

Kucinich Article Footnotes

[i] I am speculating. People die every year as a result of heat waves, tropical storms, and extreme weather. If we keep taking resources from fossil fuels we will totally destroy the environment, and then we would all die. The EPA and the DOE both state the need for renewable fuels in the coming century, and Cannabis Hemp will provide us with all our energy needs[ii] According to the Sierra Club

[iii] Department of Energy Annual Review 1999

[iv]Chris Conrad, Lifeline to the Future

[v]Department of Energy Annual Review 1999

[vi]39% according to The State of the Nation's Ecosystems, www.us-ecosystems.org/forests/index.html

[vii]According to the Sierra Club

[ix]55 million cubic metres per year every year, and a 40% increase by 2040, US Forestry Service

[x]According to the Environmental Protection Agency

[xi]According to the Environmental Protection Agency

[xii]According to the Environmental Protection Agency

TOXIC CHEMICALS and CANNABIS HEMP:
EXTINCTION and CANNABIS HEMP: PLASTIC and CANNABIS HEMP:
ECONOMY and CANNABIS HEMP:
WORLD HUNGER and CANNABIS HEMP:
THIRD WORLD DEVELOPMENT and CANNABIS HEMP:
CORAL REEFS and CANNABIS HEMP:
POACHING and CANNABIS HEMP:
THE AGRARIAN SOCIETY and CANNABIS HEMP

"THE WHOLE ECOSYSTEM IS IN JEOPARDY."

CONGRESSMAN DENNIS KUCINICH

THE EARTH'S WHOLE ECOSYSTEM IS IN JEOPARDY.

THE LAWS AGAINST THE HEMP PLANT TOOK AWAY A SUSTAINABLE WAY OF LIFE IN 1937 AND WE'VE BEEN SUFFERING EVER SINCE. YOU CANNOT TAKE THE NUMBER-ONE PLANT RESOURCE OUT OF THE ECOSYSTEM AND EXPECT ANYTHING BUT DISASTER.

HEMP MUST BE RETURNED TO THE PEOPLE FREE AND CLEAR.

Increased use of renewable energy sources, including biofuels and energy efficiency will help reduce emissions, protecting future generations from this global threat.

Modern Japanese computer controlled sailing ship, 1981 National Geographics on Energy. Six thousand or more years ago an ancestor of ours stood up on their raft held up some palm fronds and discovered sailing. We gave up sailing for fossil fuels and now we have to start over. Now its sail or the planet may die. We must start erring on the side of survival until we get a sustainable lifestyle for the planet's people.

As both oil wars in the Middle East clearly showed us, there is <u>no</u> energy secure future with fossil fuel. We now import about 70% of our oil, *money sent out of the country.* Yet this is exactly what the current administration calls its national energy strategy. More oil, more nuclear power, more oil drilling in environmentally sensitive areas [now the Artic].

HARNESSING THE WIND

IT LACKS halyards, sheets, shrouds, and most of the rigging that ships used in the past age of commercial sail. Yet the Japanese "Shin Aitoku Maru" (above) is the prototype of what may be a new fleet of sail-aided cargo ships. The incentive:

As every sailor and kite flier knows, winds constantly shift in direction and vary in speed. Although utilities would have to learn to accommodate the variable outputs of wind machines, researchers believe wind power could become a continually useful source of energy.

Its availability depends on geography. Wind generators are most practical in the Great Plains, in mountains, and along certain coastal areas. One scheme uses 200-foot-high tetrahedral wings moving around a circular track on

50% less fuel
with HEMP canvas sails.
Japanese Ship - 1981

Here we are in the Information Age, still trying to get it right and failing. There will be no provision in this energy bill for growing hemp. [In 2006, the California Legislature has passed a Hemp Farming Bill, the Governor vetoed the bill.] However, North Dakota, Hawaii and other States may beat California to the hemp prize.

We as a nation, we as farmers, have been cheated and lied to by our government. The premier resource plant in the world is illegal here - prohibited. Your energy bill may be double now what it could be with hemp. Our economy could leap forward with this new direction, along with recycling, insulation in buildings, better mileage in alcohol or fuel cell cars, more wind and solar, more alternative energy research, and we would be on the way to energy self-sufficiency. Under present conditions, because of the bogus war on hemp by the Bush Administration [now Clinton] [now Bush], scientists and bureaucrats are afraid to talk, afraid to admit that hemp could indeed save us from the synthetic society and dead end fossil fuel usage.

In 2006 Brazil is said to be two years away from energy independence using ethanol from sugarcane for fuel.

In my opinion, we will use Cannabis hemp/marijuana in a few years because we will be forced to do so by acid rain and greenhouse gasses. WHY WAIT? WHY WAIT!? Hemp is our natural fuel source. Hemp is our paper source, which will enable our forests to recover and help remove greenhouse gasses. Hemp is a multi-trillion dollar resource that can be grown at home -and on and on.

I don't get it. You, through our government, are telling me we can't use this resource because it might get me high. What kind of nonsense is that? I believe it is my right to do in my homestead as I please with hemp in pursuit of my own happiness -my inalienable right. Besides I already have the constitutional right to get high on drugs (any alcohol product, coffee, tobacco, etc.). Are we really afraid that my brain will fry like an egg in a pan, and other such lies. Somehow after smoking Cannabis hemp/marijuana for 40 years daily, I don't think it is my brain that is fried, it is our earth that is fried - fried by generations of fossil fuel (coal and oil and natural gas) burning, fried by nuclear radiation and nuclear weapons testing and nuclear power production, fried by millions of pounds of poisons used each year on our food and fiber crops, and fried by the unnecessary destruction of 70% of our forests since 1937 for paper.

GLOBAL WARMING

IS AFFECTED BY OUR FUEL CHOICES

GUESS WHAT THE CHOICES ARE?

HEMP BIO FUELS VS. TOXIC FUELS

LEGALIZE NATURE & RE-HEMP THE EARTH

I know the plant <u>Cannabis sativa</u>; I grow the plant. I want to say to all those in positions of power or persuasion in our government and elsewhere: Please do not wait one more minute to free Cannabis hemp/marijuana. If you wait, an entire year will be lost. Then a decade will be lost. [A decade has been lost since the writing of this paper]. We do not have a lot of years to waste. Do it for me, do it for our country, and do it for our planet earth. Free Hemp.

This patch suggests the question, fossil fuel or hemp?

More about hemp: Hemp's per acre output of fuel is about 10 times more than corn, at less cost than corn, and with less environmental damage than corn. "Hemp is a hearty plant that squeezes out weeds and pests, without the heavy fertilization that corn, cotton, tobacco, and other crops need. Hemp is resistant to many insects, reducing the need for chemical pesticides." - (BACH)

SCIENTIFIC AMERICAN, November, 1989. Article: "The Case for Methanol." Methanol-fueled car could integrate various features to attain higher efficiency and generate fewer emissions than a conventional gasoline-fueled car.

Hemp will produce cleaner air and reduce greenhouse gases. When biomass fuel burns, it produces CO_2 (the major cause of the greenhouse effect), the same as fossil fuel; but during the growth cycle of the plant, photosynthesis removes as much CO_2 from the air as burning the biomass adds, so hemp actually cleans the atmosphere. With the first cycle there is no further loading to the atmosphere. At this time the U.S. has not signed an international treaty to reduce greenhouse gases to 1990 levels. Maybe the Bush administration [now Clinton] [now Bush] hasn't heard about hemp, maybe you should write him a short note before he leaves, mention time is important. When biomass (hemp) is used for other more permanent applications, say a library book that will last 1500 years, and then can be recycled seven times, or building materials in a home (I never thought what it might do to the price of a home), potential greenhouse carbon is tied up and does not go back into the atmosphere. AMAZINGLY, WITH HEMP, THE FOSSIL FUELS BURNED AND POLLUTING OUR ATMOSPHERE ARE AVAILABLE ONCE AGAIN AS A RESOURCE, UNTIL A FAVORABLE CO_2 LEVEL IS REACHED.

Energy Farming In America
By: Lynn Osburn

**A practical answer to America's
farming, energy and environmental crises.**

On June 12,1989, President Bush addressed his campaign promises to deal with the pollution problems long facing the United States.

He unveiled an ambitious plan to remove smog from California and the nation's most populous cities, as well as efforts to reduce acid rain pollution. Bush recommended auto makers be required to make methanol-powered cars for use in nine urban areas plagued by air pollution. Methanol is the simplest form of primary alcohol and is commonly called wood alcohol.

About 6% of contiguous United States land area put into cultivation for biomass could supply all current demands for oil and gas without adding any net carbon dioxide to the atmosphere.

Bush called methanol "home-grown energy for America." He further proposed a 10 million ton reduction in sulfur dioxide emissions from coal-burning power plants; that's a 50% reduction over present standards. Sulfur dioxide is a major cause of acid rain, which kills 50,000 Americans and 5,000-10,000 Canadians yearly. (Brookhaven National Laboratory 1986)

William Reilly, chief of the Environmental Protection Agency, at a briefing before Bush's speech, estimated the cost of the plan would be between S14 billion and $19 billion a year after its full implementation at the turn of the century. Bush said, "Too many Americans continue to breathe dirty air, and political paralysis has plagued further progress against air pollution. We've seen enough of this stalemate. It's time to clear the air." Political paralysis seems to be a dominant trait in Washington in any given decade, but what did he mean by "stalemate?"

The root of this "stalemate" can be found in the concept of world energy resources. The industrial world currently runs on fossil fuel: natural gas, oil, and coal. Fossil fuel resources are non-renewable, being the end product of eons of natural decomposition of Earth's ancient biomass. Fossil fuels contain sulfur, which is the source of many of the aggravating environmental pollution problems threatening America.

96

Removing sulfur compounds from fossil fuels is a major expense to the energy producers. Also, burning fossil fuels releases "ancient" carbon dioxide, produced by primeval plant life eons ago, into the atmosphere causing the air we breathe to be over-burdened with CO_2 increasing the danger of global warming and the greenhouse effect.

In the late 1800s, the fledgling petroleum industry aggressively competed with the established biomass-based energy industry in a effort to gain control of world energy production and distribution. Fossil fuel producers succeeded in their campaign to dominate energy production by making fuels and chemical feedstocks at lower prices than could be produced from biomass conversion. Now the pendulum is swinging against them.

It is likely that peak oil and gas production in the coterminous United States has been reached. The bulk total production of roughly 80% will be reached by the year 2000. Peak world production will occur about the same year.

The situation for recoverable coal, world wide, is more favorable. Peak production is estimated to happen shortly after the 2100. However, increasing numbers of Americans are unwilling to accept the escalating costs of environmental pollution and destruction associated with coal-fired power plant smokestack emissions and the land destruction resulting from coal mining.

As the energy crop grows it takes in CO_2 from the air; when it is burned the CO_2 is returned to the air, creating a balanced system.

If the pollution problems inherent with fossil fuel use are solved, the dollars and cents cost of this form of energy will continue to rise due to the dwindling availability of this non-renewable world resource. On the other hand, the dollar cost of energy production from biomass conversion will remain relatively constant because the world biomass resource is renewable on a yearly basis.

The point where the cost of producing energy from fossil fuels exceeds the cost of biomass fuels has been reached. With a few exceptions, energy from fossil fuels will cost the American taxpayer more money than the same amount of energy supplied through biomass conversion.

Biomass as the term used to describe all biologically produced matter. World production of biomass is estimated at 146 billion metric tons a year, mostly wild plant growth. Some farm crops and trees can produce up to 20 metric tons per acre of biomass a year. Types of algae and grasses may produce 50 metric tons per year.

Dried biomass has a heating value of 5000-8000 Btu/lb, with virtually no ash or sulfur produced during combustion. About 6% of contiguous United States land area put into cultivation for biomass could supply all current demands for oil and gas. And this production would not add any net carbon dioxide to the atmosphere. (Environmental Chemistry, Stanley E. Manahan, Willard Grant Press, 1984)

For its Mission Analysis study conducted for the U.S. Department of Energy in 1979, Stanford Research Institute (SRI) chose five types of biomass materials to investigate for energy conversion: woody plants, herbaceous plants (those that do not produce persistent woody material), aquatic plants, and manure. Herbaceous plants were divided into two categories: those with low moisture content and those with high moisture content.

Biomass conversion may be conducted on two broad pathways: chemical decomposition and biological digestion.

Thermochemical decomposition can be utilized for energy conversion of all five categories of biomass materials, but low moisture herbaceous (small grain field residues) and woody (wood industry wastes, and standing vegetation not suitable for lumber) are the most suitable.

Biological processes are essentially microbic digestion and fermentation. High moisture herbaceous plants (vegetables, sugar cane, sugar beet, corn, sorghum, cotton), marine crops and manure are most suitable for biological digestion.

Anaerobic digestion produces high and intermediate Btu gasses. High Btu gas is methane. Intermediate-Btu is methane mixed with carbon monoxide and carbon dioxide. Methane can be efficiently converted into methanol.

Fermentation produces ethyl and other alcohols, but this process is too costly in terms of cultivated land use and too inefficient in terms of alcohol production to feasibly supply enough fuel alcohol to power industrial society.

Pyrolysis is the thermochemical process that converts organic materials into usable fuels with high fuel-to-feed ratios, making it the most efficient process for biomass conversion, and the method most capable of competing and eventually replacing non-renewable fossil fuel resources.

The foundation on which this will be achieved is the emerging concept of "energy farming," wherein farmers grow and harvest crops that are converted into fuels.

Pyrolysis is the technique of applying high heat to organic matter (lignocellulosic materials) in the absence of air or in reduced air. The process can produce charcoal, condensable organic liquids (pyrolytic fuel oil), non-condensable gasses, acetic acid, acetone, and methanol. The process can be adjusted to favor charcoal, pyrolytic oil, gas, or methanol production with a 95.556 fuel-to-feed efficiency.

Chemical decomposition through pyrolysis is the same technology used to refine crude fossil fuel oil and coal. Biomass conversion by pyrolysis has many environmental and economic advantages over fossil fuels, but coal and oil production dominates because costs are kept lower by various means including government protection.

Pyrolysis has been used since the dawn of civilization. If some means is applied to collect the off-gasses (smoke), the process is called wood distillation. The ancient Egyptians practiced wood distillation by collecting tars and pyroligneous acid for use in their embalming industry.

Pyrolysis of wood to produce charcoal was a major industry in the 1800s, supplying the fuel for the industrial revolution, until it was replaced by coal.

In the late 19th Century and early 20th Century wood distillation was still profitable for producing soluble tar, pitch, creosote oil, chemicals, and non-condensable gasses often used to heat boilers at the facility.

The wood distillation industry declined in the 1930s due to the advent of the petrochemical industry and its lower priced products. However, pyrolysis of wood to produce charcoal for the charcoal briquette market and activated carbon for purification systems is still practiced in the U.S.

The wood distillation industry used pyrolytic reactors in a process called destructive distillation. The operation was carried out in a fractionating column (a tall still) under high heat (from 1000-1700°F). Charcoal was the main fuel product and methanol production was about 1% to 2% of volume or 6 gallons per ton. This traditional method was replaced by the synthetic process developed in 1927.

The synthetic process utilizes a pyrolytic reactor operating as a gasifier by injecting air or pure oxygen into the reactor core to completely burn the biomass to ash. The energy contained in the biomass is released in the gasses formed. After purification the syngas, hydrogen and carbon monoxide in a 2 to 1 ratio, is altered by catalysts under high pressure and heat, to form methanol. This method will produce 100 gallons of methanol per ton of feed material.

Methanol-powered automobiles and reduced emissions from coal-fired power plants can become a reality by using biomass derived fuels. The foundation upon which this will be achieved is the emerging concept of energy farming, wherein farmers grow and harvest crops that are converted into fuels. Energy farming can save American family farms and turn the American heartland into a prosperous source of clean renewable energy production.

Pyrolysis is the most efficient process for biomass conversion into fuels that can replace all fossil fuel products... When farmers can grow hemp for biomass they will make a profit energy farming.

Universities, government agencies, and private firms have conducted studies looking into the feasibility of growing biomass at low cost to make fuels at affordable prices, but the most promising plant species was never considered because it is prohibited. Instead emphasis has centered around utilizing waste products: agricultural residues after harvest, forestry wastes from the timber and pulp wood industry, and municipal wastes. All of these combined cannot produce enough fuel to satisfy the needs of industry or the American consumer's automobile. Yet biomass conversion to fuel has been proven economically feasible in laboratory tests and by continuous operation of pilot plants in field tests since 1973.

Farmers should be encouraged to grow energy crops capable of producing 10 tons per acre in 90-120 days. The crop has to be naturally high in cellulose. It must grow in all climactic zones in America. And it should not compete with food production for the most fertile land. It could be grown in rotation with food crops or on marginal land where other crop production isn't profitable.

At congressional hearings on alternative fuels held in 1978, Dr. George T. Tsao, professor of chemical engineering and food and agricultural engineering, director of laboratory of renewable resources, Purdue University, said $30 per ton for biomass delivered to the fuel conversion plant is an adequate base price for the energy farmer. The price of $30/ton has also been suggested by other researchers.

Both Dr. Serge Gratch, director chemical sciences laboratory, Ford Motor Co. and Dr. Joseph M. Colucci, director fuels and lubricants General Motors Research Laboratories testified their companies were willing, especially Ford, to make cars that would run on methanol fuel. The scientists said it would take several years to tool up factories to make methanol powered autos. They said industry could solve the problems associated with methanol as fuel. And it would take about the same amount of time for the energy industry to build methanol production facilities.

So why don't we have methanol at the filling station? The scientists said the problem was government certification under the Clean Air Act required automobile manufacturers meet standards set by the EPA based on fuels available on a national level. Since methanol fuel standards had not been set, the car makers couldn't make the new fleet until the methanol fuel was available at the pump. This catch-22 situation continues today. Government is unwilling to subsidize pilot energy farms and biomass refinery construction because fossil fuel producers control the energy industry.

Hemp is the only biomass resource capable of making America energy independent. The government suspended marijuana prohibition during WWII. It's time to do it again.

The way to end this political stalemate is to start literally from the ground up. When farmers can grow hemp for biomass they will make a profit energy farming. Then it will not take long to get 6% of continental American land mass into cultivation for biomass fuels -- enough to replace our economy's dependence on fossil fuels. And as the energy crop grows it takes in CO_2 from the air; when it is burned the CO_2 is returned to the air, creating a balanced system. We will no longer be increasing the CO_2 content in the atmosphere. The threat of global greenhouse warming and adverse climatic change will diminish.

This energy crop can be harvested with equipment readily available. It can be "cubed" by modifying hay cubing equipment. This method condenses the bulk, reducing trucking costs from the field to the pyrolysis facility.

Sixty-eight percent of the energy in the raw biomass is contained in the charcoal and fuel oils made at the facility. The charcoal has the same heating value in Btu as coal, with virtually no sulfur to pollute the atmosphere. The pyrolytic fuel oil has similar properties to no. 2 and no. 6 fuel oil. The remaining energy is in noncondensible gases that are used to co-generate steam and electricity.

To keep costs down pyrolysis reactors need to be located within a 50 mile radius from the energy farms. This necessity will bring life back to our small towns by providing jobs locally. The pyrolysis facilities will run three shifts a day.

Charcoal and fuel oil can be "exported" from the rural small town in the agricultural community to the large metropolitan areas to fuel the giant power plants generating electricity. When these utility companies use charcoal instead of coal, the problems of acid rain will begin to disappear.

The charcoal can be transported economically by rail to all urban area power plants. The fuel oil can be transported economically by truck creating more jobs for Americans.

When this energy system is on line producing a steady supply of fuel for utility companies, it will have established itself in commerce. Then it will be more feasible to build the complex syngas systems to produce methanol from biomass, or make synthetic gasoline from methanol by adding the Mobil Co. process equipment to the gasifier.

To accomplish this goal of clean energy independence in America we must demand an end to hemp prohibition, so American farmers can grow this energy crop. Our government foolishly outlawed it in 1938.

Hemp is the world's most versatile plant. It can yield 10 tons per acre in four months. Hemp contains 80% cellulose; wood produces 60% cellulose. Hemp is drought resistant making it an ideal crop in the dry western regions of the country.

Hemp is the only biomass resource capable of making America energy independent. Remember that in 10 years, by the year 2000, America will have exhausted 80% of her petroleum reserves. Will we then go to war with the Arabs for the privilege of driving our cars; will we stripmine our land for coal and poison the air we breathe to drive our autos an additional 100 years; will we raze our forests for our energy needs?

During the Second World War, the federal government faced a real economic emergency when our supply of hemp was cut off by the Japanese. The federal government responded to the emergency by suspending marijuana prohibition. Patriotic American farmers were encouraged to apply for a license to grow hemp. They responded enthusiastically and grew 375,000 acres of hemp in 1943.

The argument against undertaking this massive hemp production effort today does not hold up to scrutiny.

Hemp grown for biomass makes very poor grade marijuana. The 20 to 40 million Americans who smoke marijuana would loath to smoke hemp grown for biomass, so no one could make a dime selling a farmers hemp biomass crop as marijuana.

It is time for the federal government to once again respond to our current economic emergency by utilizing the same procedure used in WWII to permit our farmers to grow American hemp so this mighty nation can once again become energy independent and smog free.

References:

U.S. Energy Atlas, David J. Cuff & William J. Young, Free Press/McMillan Publishing Co., NY, 1980

Progress in Biomass Conversion Vol. 1, Kyosti V. Sartanen & David Tillmall editors, Academic Press, NY, 1979

Brown's Second Alcohol Fuel Cookbook, Michael H. Brown (Senate hearing transcripts)

Environmental Chemistry, (4th edition), Stanley E. Manahan, P.W.S. Publishers, Boston, MA, 1979

Hemp for Victory, U.S. government documentary film, USDA 1942-43

Produced as a Public Service for the Business Alliance her Commerce In Hemp (BACH), Help Eliminate Marijuana Prohibition (HEMP) and the American Hemp Council.

Access Unlimited, P.O. Box 1900, Frazier Park, CA, 93225, 805/632-2644

Thanks to Lynn Osburn! Biofuel For Victory In The 21st Century

Corn, tree pulp and hemp are sources for clean-burning alcohol, methanol and methane gas. These 'bio-fuels' contain no sulfur, the pollutant that causes acid rain. Growing the fuel also produces oxygen, to balance the oxygen consumed during combustion. Engines stay cleaner and the air remains much cleaner.

Hemp may be the most profitable and productive fuel crop that can be grown in many areas of America. Hemp can produce about 1000 gallons of methanol per acre, four times as much as can be produced from trees. Fuel can be produced locally, reducing transportation costs. The production process, called biomass conversion, is safe and clean. It would create a domestic fuel industry, freeing us from Middle East oil dependency, providing jobs and keeping our currency at home.

Hemp fuel needs no taxpayer subsidies, as oil receives. The Department of Energy estimated that fuel could be produced from hemp for about 36 cents per gallon. In New South Wales, Australia the Minister of Energy told the parliament they should consider burning confiscated hemp to produce electricity. "It burns at extremely high temperature, produces a lot of power and is cheaper (and much cleaner) to burn than coal."

Hemp was the subject of a 1991 conference held in Wisconsin. One speaker pointed out our government spends $26 billion each year to pay farmers not to cultivate their land. Instead of this waste of taxpayer money, farmers could grow hemp or other fuel crops. This could completely end our dependence on foreign oil."

Distribution Systems

Both hemp bio-diesel and bio-butanol are liquid fuels and can be used in existing cars and trucks. Butanol seems like the best bet to use in existing service station pumps, as it is less corrosive than other alcohols, although ethanol fuels (85% ethanol, 15% gasoline) are being used in the Midwest.

To motivate customers to move toward this clean energy we suggest incentives and tax breaks. No more than a flat 20% tax should be charged...10% Federal, 5% each to the state and city/town where the transaction happened.

Following the lead of Sir Richard Branson and his cutting edge Virgin Airlines, the aviation industry should take a leadership role and fund conversions of airplane and jet engines.

Like the Master Hemp Growers Council, we can convene a conference of 'car and biofuel folks' interested in starting or growing their businesses to do the conversions from toxic to non toxic energy. This is a short window (maybe 5 years) business, but can generate substantial profits based on volume and fair pricing.

In essence, coordinated via the Master Hemp Growers Council, can be land use, how to grow hemp, growing supplies, and market opportunities.

MASTER HEMP
GROWERS
COUNCIL

| FARMERS AND PRESSERS | FARM SUPPLY COMPANIES | DISTRIBUTION SYSTEMS |

Hempification

Global warming, environmentally caused cancers, and the wrecked American economy, requires radical action to achieve the goal of an enriched, healthy environment. Here's a step by step overview of what needs to be done now to re-hemp the planet.

1. Open up no less than 10% of government land to grow massive quantities of hemp for bio fuel, building materials and other environmentally empowering uses. A recreational hemp tax of 20% could compensate the program to finance and provide technical assistance to family farms to create HEMP FOR VICTORY.

2. A task force of hemp masters from around the globe should conference regarding the process of converting from oil to hemp. Their conclusions should be put on a Project Manager (i.e. Microsoft) and implemented immediately.

3. Make land and financing available to family farms and ancillary businesses. Identifying growers and empowering them to develop family farmers and supportive ancillary industries like seed pressers, packagers, transportation and retailers.

4. Create a seed bank and distribute seeds like the EPA distributes fertilizer. Free Marc Emery so he can help out.

5. Publish growing tips for farmers.

6. Use ice bergs in the ocean to hydrate land and supply clean drinking water. Oil tankers who transport toxic energy can be used to harvest and deliver to shore intact ice bergs. This process has an added benefit by helping maintain the Salinization level in the ocean, which impacts on weather patterns.

7. Finance small on site hemp oil and fuel processing plants at the farms. Provide financing for the development, equipment and distribution start-ups.

8. Encourage the widespread use of Flex Fuel Conversion Kits: Converting cars to bio-fuel. Many jobs can be created doing the conversions. The nation's Biofuel Distribution System needs to go on a project manager too.

9. In our ignorance, we have outlawed nature: Remove ALL legislative limitations on the hemp plant. Find out how much we need to grow.

10. Adequately computerize with equipment and skills participants to help them maximize their acreage output and market opportunities.

Like in the days of Thomas Jefferson and George Washington, the family hemp farmer can grow on almost any land.

On line classes with camera hook-up in groups can teach small farmers how to grow hemp. A master's series of classes on How To Grow Hemp can create an abundant crop for processing into fuel and other helpful products and services.

"We now understand that cannabis fixes CO2. And, an acre of cannabis fixes five times as much CO2 as an acre of forest. It could be an enormous help in turning around our global warming situation.

The oil that's produced from the cannabis plant, Henry Ford used in his first cars. And people are again driving up and down the coast using biodiesel gasoline. It's an alternative to using dead dinosaurs as a way of getting around on the planet."

DR. WILLIAM L. COURTNEY—courtney@mcn.org
in an interview by Norman de Vall on KZXY/Z
The Politics Behind the Local News—WW.KZXY.ORG
published in THE NEW SETTLER, Issue 142, Summer 2007

Land

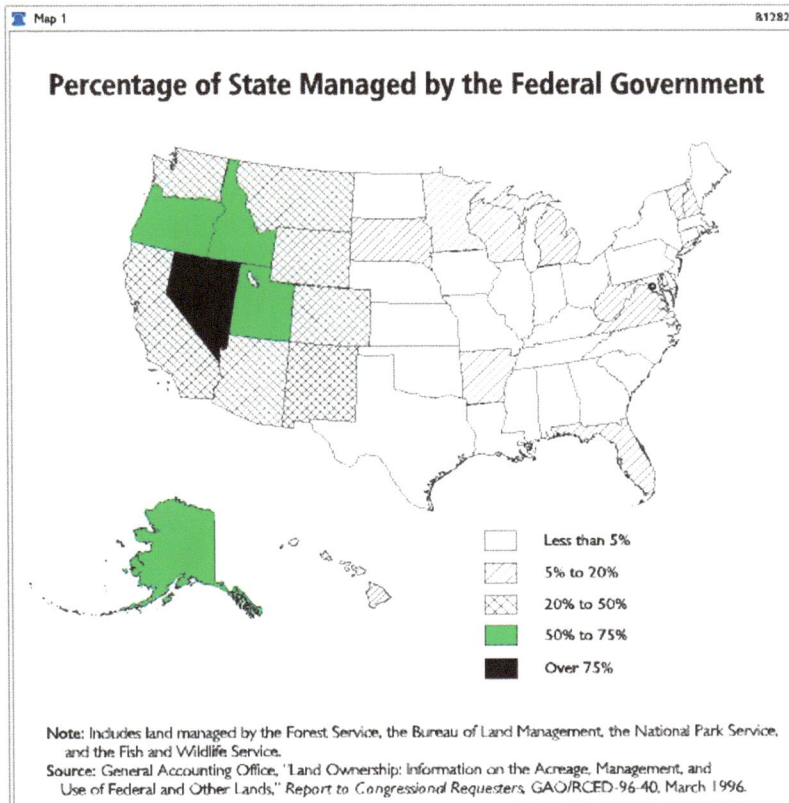

Map 1   B1282

Percentage of State Managed by the Federal Government

Legend:
- Less than 5%
- 5% to 20%
- 20% to 50%
- 50% to 75%
- Over 75%

Note: Includes land managed by the Forest Service, the Bureau of Land Management, the National Park Service, and the Fish and Wildlife Service.
Source: General Accounting Office, "Land Ownership: Information on the Acreage, Management, and Use of Federal and Other Lands," *Report to Congressional Requesters*, GAO/RCED-96-40, March 1996.

HOW MUCH LAND DOES UNCLE SAM (THE PEOPLE) OWN?

"Although the goal of preserving land for posterity is noble, the true impact of current federal land management policy should not be lost behind a cloud of good intentions. In 1996, the General Accounting Office reported that the federal government owned a staggering 650 million acres, or one-third of the land in the United States. The Bureau of Land Management, the Forest Service, the Fish and Wildlife Service, and the National Park Service manage about 95 percent of this land (approximately 618 million acres, or about 7,500 acres per employee). As of September 1994, these agencies also had obtained rights-of-use to over 3 million acres on nonfederal land through leases, agreements, permits, and easements."

WWW.HERITAGE.ORG/RESEARCH/ENERGYANDENVIRONMENT/BG1282.CFM

As demonstrated in an earlier graphic, 10% of federal land is approximately 65 million acres. In the states of California, Arizona, Nevada, New Mexico there is a lot of government owned land that is just sitting there, not being used. One can drive for hours from Bullhead City, AZ to Needles, CA to understand what I'm talking about. Miles, and miles of land that could be saving our lives.

The government needs to turn over use of these lands (and others appropriate to grow HEMP FOR VICTORY) to a Master Hemp Growers Council, to grow sufficient amounts of hemp to replace the use of polluting oil.

**CALIFORNIA HEMP MASTERS
RICHARD M. DAVIS AND R. W. AKILE
LOS ANGELES MILLION MARIJUANA MARCH, 2006**

"For when the [American] revolution took place, the people of each state became themselves sovereign; and in that character hold the absolute right to all their navigable waters, and the soils under them, for their own common use, subject only to the rights since surrendered by the Constitution to the general government."-Martin vs. Waddell (1842) 41 US (16 Pet) 367, 410

Seeds

Different strains of hemp are better for certain applications. An immediate seed bank needs to be established. Hemp seed expert Marc Emery, an international leader in the hemp movement, can help organize the seed bank to supply the best seeds for the task at hand. The definition of "best seeds" needs to come from the Master Hemp Growers Council, which should be quickly convened.

Planting

Here's where the jobs come in. In addition to folks working in the field, there are the ancillary jobs of feeding, housing, clothing and entertaining them.

Microsoft Project Manager style use of resources can help growers supply the market, pre purchased (farming co-ops) in some cases.

Water/Irrigation: There have been ice bergs the size of states breaking off, slowly melting into the ocean, changing the salt balance of the ocean. Explains some of the weather we've been having.

Harvesting large ice bergs and transporting them to open areas will for a relatively low cost, hydrate the land.

Tending Crops and Harvesting - Processing from plant to fuel.

Once the plants are grown they can be processed on site to the correct grade fuel. Hemp pellets can be used to run the power plants.

There are contracts for transporting the fuels, insuring the farm, fuel processors.

Financial Empowerment – small business and jobs – new tax source. Extra tax credits should be given as investment incentives.

Hemp
Fuel For The 21st Century Links

Biomass Sites

Biopower Basics (DOE)
www.eren.doe.gov/biopower

Energy Ideas Clearinghouse
www.energy.wsu.edu

Great Lakes Biomass Energy Program
www.cglg.org/projects/biomass

National Renewable Energy Lab.
www.nrel.gov

Oak Ridge National Laboratory
www.ornl.gov

Western Biomass Quarterly
www.westbioenergy.org

Related Associations/Orgs

American Forest & Paper Assoc.
www.afandpa.org

**California Forest
Products Commission**
www.calforests.org

California Forestry Association
www.foresthealth.org

CA Licensed Foresters Assoc.
www.clfa.org

**Independent Energy Producers
Association**
www.iepa.com

**Iowa State University,
Office of Biorenewables Progs.**
www.biorenew.iastate.edu/

Public Policy Advocates
www.ppallc.com

Quincy Library Group
www.qlg.org

Society of American Foresters
www.safnet.org

Government/Agencies

California Air Resources Board
www.arb.ca.gov

**California Department of Forestry and Fire
Protection**
www.fire.ca.gov

California Energy Commission
www.energy.ca.gov

**California Environmental Protection
Agency**
www.epa.ca.gov

California Resources Agency
www.ceres.ca.gov

**California Integrated
Waste Management Board**
www.ciwmb.ca.gov

U.S. Department of Agriculture
www.usda.gov

U.S. Department of Energy
www.doe.gov

**U.S. Dept. of the Interior/Bureau of Land
Management**
www.blm.gov

USDA Forest Service
www.fs.fed.us

10. LETTERS TO THE LOS ANGELES TIMES

BIOFUELS AND GLOBAL WARMING

SEPTEMBER 23, 1999

RE: Unilateralism Must Give Way...Opinion, Sept. 23, 1999

Maksoud's call for Internationalism and empowering the U.N. lacks a fundamental understanding of American history's lesson in government. The Continental Congress was a confederation like the U.N. under the Articles of Confederation for eleven years. The confederation was a failure at securing peace between sovereign states, as it had no powers of direct enforcement of its laws. In addition, The Continental Congress had no independent taxing powers, could not regulate interstate and foreign commerce, was ineffective in foreign affairs, had no chief executive, and had no binding court of justice.

Every one of these charges can be leveled at the United Nations Confederation today. **The solution in the 13 colonies was to hold a Constitutional Convention** that created a Federal style of government that has served as a model for 200 years. We have a model. We need a world constitutional convention to reform the United Nations into a world federal government. Imagine the preamble: We the People of the World, in Order to form a more perfect Union, establish Justice, insure domestic Tranquility, provide for the common defense, promote the general Welfare, and secure the Blessings of Liberty to ourselves and our Posterity, do ordain and establish this Constitution for the World.

The World needs government that works, arresting individuals not attacking nations, controlling multi-national corporations and removing the nuclear madness that still exists. [And isn't Global Warming a planet wide issue?] The United States is a model, not the world policeman.

Richard M. Davis, Curator, U.S.A. Hemp Museum

Gas Prices: No Cheap Fix
MARCH 1, 2000
Los Angeles Times: Letter to the Editor

We do have a choice in energy production, which most Americans including the Times choose to ignore. The Times blames our dwindling reserves, our appetites for gas, gas guzzlers, and extra miles for the higher prices. I blame our lack of energy policy, which includes the prohibition of hemp for energy, the farmer's best crop for this purpose.

In 1980, V.P. candidate George Bush stood at a fueling station pumping methanol into the tank of a car, touting the use of alternative fuels. Fuel for cars or power plants does not equal fossil oil or coal. Anything that can be made from fossil fuels can be made from biomass (biologically produced matter). In other words we could give energy dollars to American farmers, instead of being held hostage by Iraq, Venezuela, Norway, or Mexico. Let's buy our energy reserves locally. Hemp is now legal in Canada, why not here?

Richard M. Davis, Curator, U.S.A. Hemp Museum

From High Times Magazine, February, 1995

FREEDOM FIGHTER OF THE MONTH: RICHARD DAVIS
Curator of the Traveling Hemp Museum

By Bill Bridges

"Richard Davis is not new to the hemp movement. He has smoked and grown for over 27 years. In 1992 he ran for President on a hemp platform. In 1986 he ran for Congress as a pot-grower. He has displayed marijuana buds at the Capital and helped with both California's and Colorado's hemp initiatives..."

"Three years after marijuana prohibition started, you could still go down to the drugstore and buy a bottle of cannabis extract," says Davis. "No one knew that marijuana and cannabis were the same thing. Those laws took away a sustainable way of life, and we've been suffering ever since. **You cannot take the number one plant resource out of the ecosystem and expect anything but disaster...**"

"Without hemp you don't know your own history," he says. "You don't know what the potential is for saving the planet. That's why the Hemp Museum is so important. Just learn the truth."

112

Creating New Government

**DECEMBER 2, 2001—Attn: Robert Berger
Los Angeles Times: Op-Ed Commentary**

The failure of the Afghanistan government and the need for new law there also points to the need for government on a world level. By examining United States history and mistakes made in creating the world's first written constitution, we can avoid needless setbacks in drafting law on the world level.

It took our founding fathers eleven years to figure out that the Articles of Confederation would not work to govern the colonies in early America. The Continental Congress had: no taxing powers, no regulation of interstate and foreign commerce, no powers of direct enforcement of its laws, was ineffective in foreign affairs, no chief executive, no binding court of justice. They did not attempt to fix the Articles of Confederation. They called instead for a constitutional convention to draft the document that has stood the test of time, now 212 years, to govern these United States of America.

The U.S. Constitution is one model for what today is sorely overdue on our singular planet home - workable government. The United Nations has most of the above attributes of the Articles of Confederation and one question is can it be redrafted? Or can we in the spirit of our founders realize the power of starting over, and call for a World Constitutional Convention. We have been given a model that works, that can change, that arrests individual lawbreakers rather than invades states, that gives us a bill of rights against the power of government in our individual lives.

Of course the government we have is not perfect, but it offers ways to fix it through legislation, courts, and amendments. And the people of Afghanistan or the world don't need to repeat the mistakes we made in our own history in instituting government. Women were not equally represented in the writing of the document, were not given rights or the vote, were not equally represented at all levels of government. Likewise, racism and slavery were not dealt with and almost brought about the downfall of the government in civil war. Monetary and corporate questions must be carefully addressed. And religion was placed squarely outside of government.

One thing our U.S. Constitution was never equipped to do was make us the world policeman. Our assuming this role on occasion only points out the anarchy that exists due to the lack of enforceable world law and the lack of workable world government. We cannot secure the rights of individuals on a world stage without world law. We can bomb Iraq, Serbia, and Afghanistan, but cannot arrest Saddam, Milosavic, or Bin Laden. We ignore Africa and East Timor, and the dozens of other wars now underway. We ignore the call for nuclear disarmament and environmental greenhouse gas reductions, we want out of the ABM missile treaty, because we are sovereign and can get away with it, not because the earth depends on it.

The same goes for ignoring U.N. dues. We have let our own fundamentalist religious beliefs stand in the way of population reduction efforts and an equal rights amendment for women. The women of Afghanistan have suffered at the hands of religious fundamentalists and should be equally represented in the new government there. How can we recommend that for Afghanistan if we cannot see the logical extension to our own government?

Secretary of State Madeleine Albright (L.A. Times, Oct. 6, 1999) decried the low status and "appalling abuses committed against women ...including coerced abortions and sterilizations, children sold into prostitution, ritual mutilations, dowry murders and domestic violence ...(women are) exploited, discriminated against and even sold (slavery)." Then she said, "In our diplomacy we are working with others to change that."

Diplomacy and promoting the cause of women's rights with foreign aid cannot replace international law with enforcement. We have a U.N. Declaration of Human Rights, but no workable world government to back up those rights.

The most important job of world government is to end war between nations, and thereby protect individual rights, a government of the people, by the people, and for the people, not for unrestrained capitalism or multinational corporations. Our own Declaration of Independence names this core purpose of government: "We hold these Truths to be self-evident, that all Men (people) are created equal, that they are endowed by their Creator with certain unalienable Rights, that among these are Life, Liberty, and the Pursuit of Happiness - That to secure these Rights, Governments are instituted among Men (people)." As we see around the World, no individual right is secure when planes and bombs are falling, or genocide is happening, or hate based discrimination is the order of the day.

Just as it is time for new government in Afghanistan, it is time for the people of the world to sit down at the table, women and men together, in equal numbers, and write a new document of government for our common home-the planet earth. President Eisenhower said this very plainly in 1956: "There can be no peace without law." We the people of the world need a World Constitutional Convention. We have the models, and we have the minds.

Richard M. Davis

VOTEHEMP.COM
VOTE FOR HEMP RIGHTS!

Greens Don't See Forest...
Mar. 26, L.A. Times
RE: Commentary

Patrick Moore, founder of Greenspirit, ought to be ashamed of himself for another misleading and inaccurate article. This type of article makes me wonder who funds Greenspirit.

Mr. Moore says the battle for America's forests was fought 100 years ago, but failed to state the fact that sustainable forestry does not exist today and has not existed for the past 100 years.   U.S. Dept. of Agricultural Bulletin 404, 1916, showing hemp made fine paper, warned of the pressure on our forests from paper production alone. And we now have four times the population of 100 years ago.

Mr. Moore makes this speech in a previous article on trees and wood: "But now, it is so trendy to be opposed to cutting trees that many people find it possible to ignore the absolute necessity of using wood in their everyday lives. Many seem willing to forget that wood is, without question, the most renewable and environmentally friendly of all materials used to build our civilization. Wood is the material embodiment of solar energy, created by photosynthesis in a factory called the forest, and whether we like it or not, wood can only be obtained from trees."

His speech is misleading and its conclusion false. Mr. Moore should read Bulletin 404. It clearly states that four times as much wood can be obtained from a woody herbaceous shrub called hemp than from any tree in the same period. And I'm sitting here looking at a sample of medium density fiberboard from hemp that is stronger than that from trees and typing on a computer that could be made from hemp plastic. Hemp fuels could replace fossil fuels. And hemp can be grown on a "factory" called the family farm.

He equates tree farms to forests. Forests are not factories; they are ecosystems, full of natural relationships of life that are destroyed by logging. Wood may be a necessity of everyday life, but so are living trees. And saving living trees and using hemp and other plant sources for wood is the way to reverse the greenhouse gas buildup, because all plants absorb CO_2 and give off oxygen.

Now you know, Mr. Moore, which wood (hemp) is, without question, the most renewable and environmentally friendly, which wood (hemp) is the embodiment of solar energy, whether we like it or not. The Green Party endorses hemp. For years we have tried to get the major environmental groups to acknowledge and work for hemp -Greenpeace, the Sierra Club. Now you know, Mr. Moore. Now Greenspirit knows. What are you going to do about it?   The American farmers would love to hear that the prohibition against growing hemp has been lifted.

Richard M. Davis, Curator, USA Hemp Museum

Energy, Fuel For U.S.
JUNE 13, 2003

If the U.S. is advocating a return to nuclear power, offshore drilling for oil and gas, we are in a crisis now. How are we to respond as a nation? We will not respond in a sane fashion if all the alternatives are not presented to the people by the press. The people need to know that clean burning fuel can be grown by California farmers in the form of ethanol, methanol, or bio-diesel.

Hemp is ten times as productive as corn for ethanol, with less water, pesticides and fertilizers. Ethanol from corn is now imported into California from the mid-west to help clean gasoline emissions. Politics of the drug war have prevented the growing of hemp. Those policies need to be reversed immediately. Homegrown energy will help keep us out of oil wars, help us reverse global warming and keep our energy dollars at home.

Richard M. Davis, Curator

NAIHC
NORTH AMERICAN INDUSTRIAL HEMP COUNCIL
P.O. BOX 259329, MADISON, WI 53725-9329
(608) 258-0243

CHAIR@NAIHC.ORG
WWW.NAIHC.ORG

HEMP HERO LORNA MILNE

Senator for Ontario, Canada

On June 19, 1996, Senator Milne successfully proposed an amendment to the government's drug legislation, Bill C-8, authorizing the cultivation of hemp in Canada. (From her website: Global Warming - Impact on the Arctic)

Hon. Lorna Milne: Honourable senators, I rise this afternoon to highlight a recent report commissioned by the Arctic Council, a report entitled *The Arctic Climate Impact Assessment.* The Arctic Council is a group of national governments and Aboriginal organizations working together to study issues that have an impact on the world's Arctic region. The study focused on the impact of global warming on the Arctic region, and I can tell honourable senators that the news is not good. Some of us already live with the problem.

The key finding in this report is confirmation that global warming is hitting the Arctic earlier and harder than most of the rest of the world. The models show that temperature will increase in our Arctic at double the rate that it will in the rest of the world. The specific results of this increase in temperature will be significant, and I urge all honourable senators to reflect on how some of these changes will affect their communities and, indeed, the planet's biodiversity.

Vegetation zones are moving northward as a result of the warming. Left unchecked, this will likely lead to frequent forest fires and increased insect outbreaks. We have already seen this very dramatically in the province of British Columbia.

The range in distribution of animal species will also shift. The result will be a decrease in the habitat area for many northern plant and animal species and could bring new natural predators to the region. Consequently, there is the potential to push some species toward extinction, including polar bears, caribou and some seabirds.

WWW.SEN.PARL.GC.CA/LMILNE

Coastal communities also face significant damage from unchecked global warming. Changes in the heights of tides and ocean currents will have an impact on both erosion and flooding. This has the potential to threaten many Canadian communities in the north, most of them Aboriginal.

Finally, although the ozone layer issue has been in some part addressed, the depleted ozone layer is still a serious problem in the Arctic. Global warming is exacerbating the historic damage to the ozone layer, and scientists predict that it will take decades before the layer over the Arctic is fully healed. Young Canadians in the North now receive ultraviolet radiation doses at least 30 per cent higher than any previous generation. This will probably have a significant impact on cancer rates in the North as the years go on.

Honourable senators, these are just a few of the problems that face Canada's Arctic if global warming is left unchecked. I do not have to tell honourable senators of the dire consequences of an increase in carbon dioxide in the atmosphere when the permafrost increases its rate of thawing.

I hope that the Senate will continue the work that has been started by Senator Banks and the Standing Senate Committee on Energy, the Environment and Natural Resources in the last Parliament to find ways to stop or at least slow down global warming. I strongly urge the federal government to implement a comprehensive program to protect Canada's Arctic.

Industrial Hemp Links

The Canadian Industrial Hemp Council
WWW.CINEVISION.COM/CIHC

Health Canada's Industrial Hemp Fact Sheet
www.hc-sc.gc.ca/english/media/releases/1998/hemp-e.htm

Kenex - A hemp research and production company
WWW.KENEX.COM

The North American Industrial Hemp Council
WWW.NAIHC.ORG

Canada has a 10 year head start in the development
of the hemp industry and products.

11. READINGS:
HEMP AND GLOBAL WARMING

HEMP BIOMASS FOR ENERGY, By Tim Castleman, (2001) www.lulu.com
GLOBAL WARMING, Greenpeace (1990), edited by Jeremy Leggett.
AN INCONVENIENT TRUTH: The Planetary Emergency of Global Warming and What We Can Do About It. (2006) By Al Gore.
WATER: The Power, Promise, and Turmoil of North America's Fresh Water. (1993) National Geographic Special Edition.)

THE SOLUTION IS YOU:

AN ACTIVIST GUIDE

BY: LAURIE DAVID

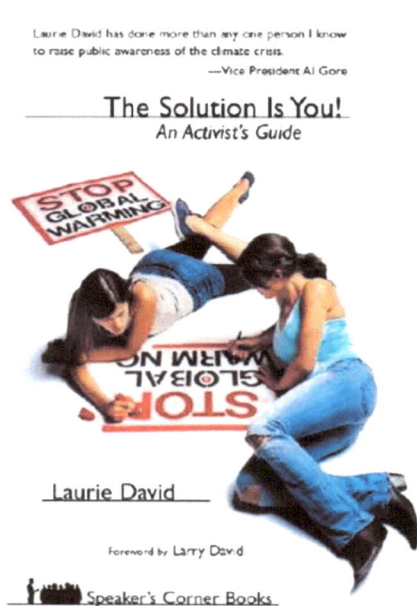

Laurie David has done more than any one person I know to raise public awareness of the climate crisis.
—Vice President Al Gore

The Solution Is You!
An Activist's Guide

Laurie David

Foreword by Larry David

Speaker's Corner Books

Laurie David
Photo: Tierney Gearon

WWW.LAURIEDAVID.COM

HEMP HERO CHRIS CONRAD

Chris Conrad is an internationally respected authority on cannabis, industrial hemp, medical marijuana, cultivation, yields and cannabis culture. He was also editor and designer of the first modern edition of <u>The Emperor Wears No Clothes</u>. Chris and his wife Mikki Norris are early pioneers and exemplary activists in the modern hemp movement. Their contributions are too many to list in this book.

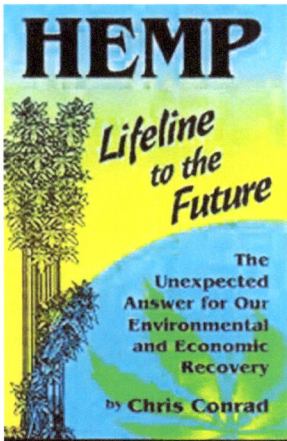

FROM CHRIS CONRAD'S BOOK

HEMP LIFELINE TO THE FUTURE

Chapter 10, Energy Independence & Security page 108, 1994 Edition: "During the Second World War, the head of the Hemp For Victory program explained that hemp was again powering its own mechanical processing and generating a 50 percent energy surplus. "Fiber is obtained from the stems of the plant, cannabis sativa. All of the factories use the hurd to fire the huge boilers which provide heat for drying and power to operate the machines. Fuel costs are eliminated through this ingenious procedure.[21] Imagine the potential now that better technology and cogeneration power are available."

WWW.CHRISCONRAD.COM

21 Only 20% of the hurd was used for fuel, disposing of the rest was a problem for the plant. Plans and descriptions of Hemp processing factories follow. Sackett & Hobbs. Hemp; A War Crop, Mason & Hanger Co. New York NY. 1942

Scientific American

November, 1989

Methanol-fueled car could integrate various features to attain higher efficiency and generate fewer emissions than a conventional gasoline-fueled car.

**Transit Bus Operation
with Methanol Fuel**

**School Bus
Demonstration Project**

California Energy Commission

122

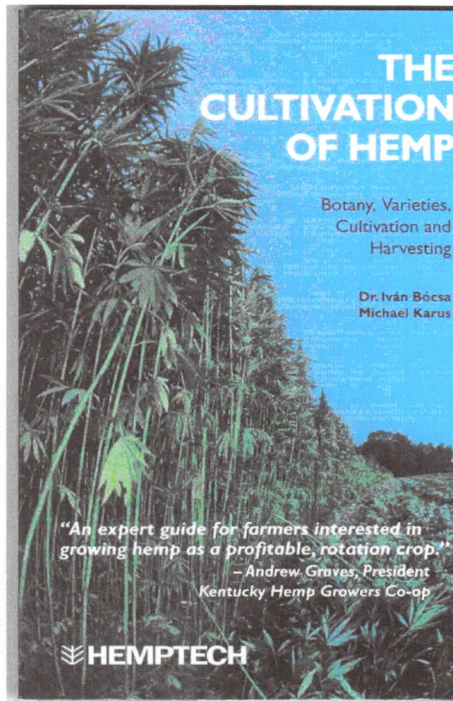

THE
CULTIVATION
OF HEMP

Botany, Varieties,
Cultivation and
Harvesting

Dr. Iván Bócsa
Michael Karus

"An expert guide for farmers interested in growing hemp as a profitable, rotation crop."
— Andrew Graves, President Kentucky Hemp Growers Co-op

HEMPTECH

"The Cultivation of Hemp: Botany, Varieties, Cultivation and Harvesting"

Dr. Ivan Bocsa and Michael Karus. 1998. HEMPTECH. Sebastopol, CA

CONTENTS:

1. Hemp's Historical Significance.

2. Hemp Cultivation Today.

3. Hemp's Origin and Botany.

4. Breeding of Hemp Varieties.

5. Hemp Cultivation.

6. Harvesting.

7. Hempseed Cultivation.

8. An Ecological Evaluation of Hemp Cultivation.

9. New Uses for Hemp in Western Europe.

THE MOTHER EARTH NEWS

BIOMASS CAR

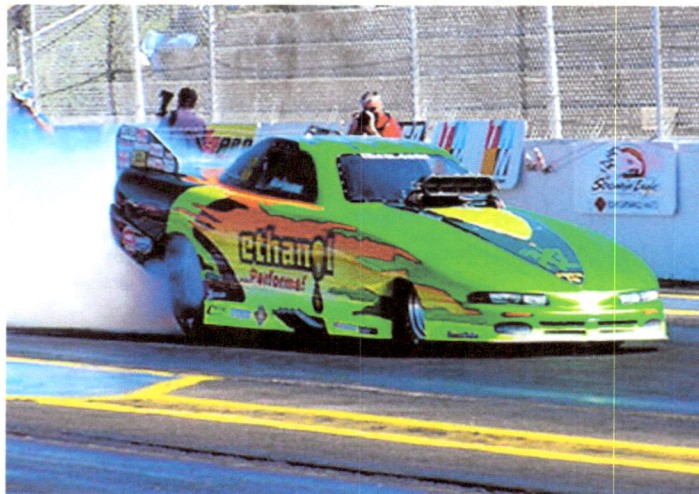

WWW.EVWORLD.COM

"PHOTO CAPTION: Five-time IHRA funny car champion Mark Thomas' Dodge Avenger runs on 'straight ethanol' and regularly turns in speeds of more than 200 mph. Ethanol is a renewable fuel that can be made not only from corn, but plant residue; it is carbon neutral, adding no additional greenhouse gases to the earth's atmosphere."

12. HEMP LEGISLATION
California's Industrial Hemp Bill

The following California Assembly Bill Number 1147 has passed the California Legislature and was vetoed by Governor Schwarzenegger.   This was a no brainer, a vote for survival.

BILL NUMBER: AB 1147
AMENDED BILL TEXT
AMENDED IN ASSEMBLY
MARCH 30, 2005
Introduced by:
Assembly Member Mark Leno

**Assembly Member
Mark Leno**

FEBRUARY 22, 2005
An act to add Division 26 (commencing with Section 81100) to the Food and Agricultural Code, relating to industrial hemp.

LEGISLATIVE COUNSEL'S DIGEST

AB 1147, as amended, Leno. Industrial hemp: license for commercial purposes.

(1) Existing law contained in the Food and Agricultural Code does not authorize the production or utilization of industrial hemp in this state. The Food and Agricultural Code provides that a violation of any of its provisions is, in general, a misdemeanor.

This bill would provide that any person desiring to grow industrial hemp, *as defined,* for Commercial purposes or *to* operate as a primary processor of viable hemp seed into commercial, nonviable seed derivatives shall apply to the Department of Food and Agriculture for a license; the bill would require any licensee to meet specified conditions. The bill would provide for the assessment of a fee on license applicants and for research by the University of California on industrial hemp, as specified. By creating new crimes, this bill would impose a state-mandated local program upon local governments.

(2) The California Constitution requires the state to reimburse local agencies and school districts for certain costs mandated by the state. Statutory provisions establish procedures for making that reimbursement.

This bill would provide that no reimbursement is required by this act for a specified reason.

Vote: majority. Appropriation: no. Fiscal committee: yes. State-mandated local program: yes.

THE PEOPLE OF THE STATE OF CALIFORNIA DO ENACT AS FOLLOWS:

SECTION 1. Division 26 (commencing with Section 81100) is added to the Food and Agricultural Code, to read:

DIVISION 26. INDUSTRIAL HEMP

81100. Unless otherwise provided or the context otherwise requires, the definitions in this section shall govern the construction of this division:

(a) "Secretary" means the Secretary of Food and Agriculture, or the secretary's designee.

(b) "Department" means the Department of Food and Agriculture.

(c) "Industrial hemp" is generally an oilseed and fiber crop that includes all parts and varieties of the plant Cannabis Sativa L, having no more than three-tenths of one percent tetrahydrocannabinol contained in its dried flowering tops; and that is grown wholly within this state from indigenous instate seed stock exclusively for the purpose of producing sterilized stalk, fiber, and seed elements and products thereof.

(d) "Tetrahydrocannabinol" or "THC" means the natural or synthetic equivalents of the substances contained in the plant, or in the resinous extractives of, cannabis, or any synthetic substances, compounds, salts, or derivatives of the plant or chemicals and their isomers with similar chemical structure and pharmacological activity.

81102. (a) Any person desiring to (1) grow industrial hemp for commercial purposes; or (2) operate as a primary processor of viable hemp seed into commercial nonviable seed derivatives shall apply to the Department of Food and Agriculture for a license on a form prescribed by the department.

(b) The department shall adopt regulations establishing criteria for the issuance of licenses, which criteria shall include, but need not be limited to, the following:

(1) Permitholders are not authorized to sell or trade viable hemp seed outside of California.

(2) Licenses shall be subject to renewal after two years.

(3) Background and qualifications of the applicant must be submitted, which shall include a complete state and federal summary criminal history check, at the applicant's expense.

(4) No person with a prior criminal conviction shall be eligible for a license.

81104. Every licensee under this division shall be subject to the following conditions:

(a) (1) Each licensee shall file with the Department of Food and Agriculture documentation indicating that the seeds planted were of a type and variety certified to have no more than three-tenths of one percent tetrahydrocannabinol and a copy of any contract to grow industrial hemp.

(2) The department shall adopt rules that provide for testing industrial hemp during growth for tetrahydrocannabinol levels and for supervision of the crop during growth and harvest.

(b) No licensee may remove from its operation any item or element other than mature stalks, fiber, or viable seed for sale, distribution, or introduction into the commerce of this state.

(c) A licensee may sell or distribute mature stalks, fiber, or viable seed only to a primary processor licensed under this division.

(d) Each person licensed to grow industrial hemp shall notify the Department of Food and Agriculture of the sale or distribution of any industrial hemp seed or stalk grown by the licensee, and of the names of the licensed persons to whom any viable hemp seed was sold or distributed.

(e) Each person licensed as a primary processor shall render each seed into a nonviable seed derivative, including, but not limited to, oil, nut, or powder.

(f) Each person licensed as a primary processor shall test the tetrahydrocannabinol (THC) levels of any derivative product using a laboratory registered with the federal Drug Enforcement Administration and shall report the results of those tests to the Department of Food and Agriculture, in a form and on a schedule set forth in regulations adopted by the department.

(1) In every case, for hemp oil products grown in this state, the trace tetrahydrocannabinol content shall not exceed more than five parts per million (ppm) of tetrahydrocannabinol.

(2) In every case, for hemp nut products grown in this state, the trace tetrahydrocannabinol content shall not exceed more than 1.5 parts per million (ppm) of tetrahydrocannabinol.

81108. To provide sufficient funds to pay all costs associated with monitoring and testing in the state, the Department of Food and Agriculture shall assess each applicant a fee in an amount determined by the department to cover those costs.

81110. The University of California shall be authorized to conduct research relating to the production and processing of industrial hemp, as follows:

(a) One of the purposes of the research shall be the development and dissemination of technology important to the production and utilization of commercial crop and livestock enterprises.

(b) The research shall provide for the enhancement of the quality of life, sustainability of production, and protection of the environment.

(c) As a part of this research, the university may collect feral hemp seed stock and develop appropriate adapted strains of industrial hemp which contain less than three-tenths of one percent tetrahydrocannabinol in the dried flowering tops.

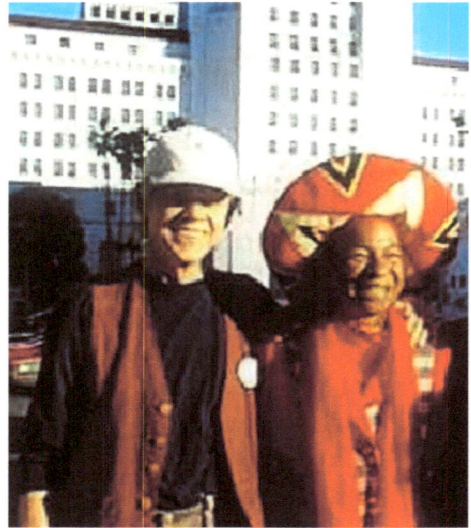

RICHARD M. DAVIS AND SISTER SOMAYAH KAMBUI WINNING THE RIGHT TO GROW HEMP IN CALIFORNIA

(d) The university shall report its findings to the Department of Food and Agriculture.

SEC. 2.

No reimbursement is required by this act pursuant to Section 6 of Article XIII B of the California Constitution because the only costs that may be incurred by a local agency or school district will be incurred because this act creates a new crime or infraction, eliminates a crime or infraction, or changes the penalty for a crime or infraction, within the meaning of Section 17556 of the Government Code, or changes the definition of a crime within the meaning of Section 6 of Article XIII B of the California Constitution.

The United Nations Law
On Industrial Hemp

The U.N. Single Convention on Narcotic Drugs, 1961, Article 28, exempts the planting of industrial hemp from prohibition. This is the way many countries such as Canada are breaching the prohibition mentality (See Hemp Hero Lorna Milne, page 118).

According to the U.S. Government classification of Cannabis, the Canadians and Europeans are growing huge fields of marijuana. This is a treaty that was signed by the United States.

Countries Growing Industrial Hemp Today

The U.S. is the only industrialized nation in the world that does not recognize the value of industrial hemp and permit its production. Below is a list of other countries that are more rational when it comes to hemp policy.

AUSTRALIA began research trials in Tasmania in 1995. Victoria commercial production since 1998. New South Wales has research. In 2002, Queensland began production. Western Australia licensed crops in 2004.

AUSTRIA has a hemp industry including production of hemp seed oil, medicinals and Hanf magazine.

CANADA started to license research crops in 1994. In addition to crops for fiber, one seed crop was licensed in 1995. Many acres were planted in 1997. Licenses for commercial agriculture saw thousands of acres planted in 1998. 30,000 acres were planted in 1999. In 2000, due to speculative investing, 12,250 acres were sown. In 2001, 92 farmers grew 3,250 acres. A number of Canadian farmers are now growing organically-certified hemp crops (6,000 acres in 2003 and 8,500 acres in 2004, yielding almost four million pounds of seed).

CHILE has grown hemp in the recent past for seed oil production.

CHINA is the largest exporter of hemp textiles. The fabrics are of excellent quality. Medium density fiber board is also now available. The Chinese word for hemp is "ma."

DENMARK planted its first modern hemp trial crops in 1997. The country is committed to utilizing organic methods.

FINLAND had a resurgence of hemp in 1995 with several small test plots. A seed variety for northern climates was developed called Finola, previously know by the breeder code "FIN-314." In 2003, Finola was accepted to the EU list of subsidized hemp cultivars. Hemp has never been prohibited in Finland. The Finnish word for hemp is "hamppu."

FRANCE has never prohibited hemp and harvested 10,000 tons of fiber in 1994. France is a source of low-THC-producing hemp seed for other countries. France exports high quality hemp oil to the U.S. The French word for hemp is "chanvre."

GERMANY banned hemp in 1982, but research began again in 1992, and many technologies and products are now being developed, as the ban was lifted on growing hemp in November, 1995. Food, clothes and paper are also being made from imported raw materials. Mercedes and BMW use hemp fiber for composites in door panels, dashboards, etc. The German word for hemp is "hanf."

GREAT BRITAIN lifted hemp prohibition in 1993. Animal bedding, paper and textiles markets have been developed. A government grant was given to develop new markets for natural fibers. 4,000 acres were grown in 1994. Subsidies of 230 British pounds per acre are given by the government to farmers for growing hemp.

HUNGARY is rebuilding their hemp industry, and is one of the biggest exporters of hemp cordage, rugs and fabric to the U.S. They also export hemp seed, paper and fiberboard. The Hungarian word for hemp is "kender."

INDIA has stands of naturalized Cannabis and uses it for cordage, textiles and seed.

ITALY has invested in the resurgence of hemp, especially for textile production. 1,000 acres were planted for fiber in 2002. Giorgio Armani grows its own hemp for specialized textiles.

JAPAN has a rich religious tradition involving hemp, and custom requires that the Emperor and Shinto priests wear hemp garments in certain ceremonies, so there are small plots maintained for these purposes. Traditional spice mixes also include hemp seed. Japan supports a thriving retail market for a variety of hemp products. The Japanese word for hemp is "asa."

NETHERLANDS is conducting a four-year study to evaluate and test hemp for paper, and is developing specialized processing equipment. Seed breeders are developing new strains of low-THC varieties. The Dutch word for hemp is "hennep."

NEW ZEALAND started hemp trials in 2001. Various cultivars are being planted in the north and south islands.

POLAND currently grows hemp for fabric and cordage and manufactures hemp particle board. They have demonstrated the benefits of using hemp to cleanse soils contaminated by heavy metals. The Polish word for hemp is "konopij."

ROMANIA is the largest commercial producer of hemp in Europe. 1993 acreage was 40,000 acres. Some of it is exported to Hungary for processing. They also export hemp to Western Europe and the U.S. The Romanian word for hemp is "cinepa."

RUSSIA maintains the largest hemp germplasm collection in the world at the N.I. Vavilov Scientific Research Institute of Plant Industry (VIR) in St. Petersburg. They are in need of funding to maintain and support the collection. The Russian word for hemp is "konoplya."

SLOVENIA grows hemp and manufactures currency paper.

SPAIN has never prohibited hemp, produces rope and textiles, and exports hemp pulp for paper. The Spanish word for hemp is "cañamo."

SWITZERLAND is a producer of hemp and hosts one of the largest hemp trade events, Cannatrade.

TURKEY has grown hemp for 2,800 years for rope, caulking, birdseed, paper and fuel. The Turkish word for hemp is "kendir."

UKRAINE, EGYPT, KOREA, PORTUGAL and THAILAND also produce hemp.

UNITED STATES granted the first hemp permit in over 40 years to Hawaii for an experimental quarter-acre plot in 1999. The license was renewed, but the project has since been closed due to DEA stalling tactics and related funding problems. Importers and manufacturers have thrived using imported raw materials. 22 states have introduced legislation, including VT, HI, ND, MT, MN, IL, VA, NM, CA, AR, KY, MD, WV and ME, addressing support, research or cultivation with bills or resolutions. The National Conference of State Legislators (NCSL) has endorsed industrial hemp for years.

Bibliography

Chris Conrad, "Hemp: Lifeline to the Future"
Jack Frazier, "The Great American Hemp Industry"
Hemptech, "Industrial Hemp" and "Hemp Horizons"

Source: The Hemp Industries Association
http://www.thehia.org/facts.html#Countries

The Marijuana Tax Act of 1937

Full Text of the Marihuana Tax Act as passed in 1937
Introduction (in italics) by David Solomon

The popular and therapeutic uses of hemp preparations are not categorically prohibited by the provisions of the Marihuana Tax Act of 1937. The apparent purpose of the Act is to levy a token tax of approximately one dollar on all buyers, sellers, importers, growers, physicians, veterinarians, and any other persons who deal in marijuana commercially, prescribe it professionally, or possess it.

The deceptive nature of that apparent purpose begins to come into focus when the reader reaches the penalty provisions of the Act: five years' imprisonment, a $2,000 fine, or both seem rather excessive for evading a sum (provided for by the purchase of a Treasury Department tax stamp) that, even if collected, would produce only a minute amount of government revenue. (Fines and jail sentences were further increased to the point of the cruel and unusual in subsequent federal drug legislation that incorporated the Marijuana Tax Act. It is now possible under the later version of the Act to draw a life sentence for selling just one marihuana cigarette to a minor.) One might wonder, too, why a small clause, amounting to an open-ended catchall provision, was inserted into the Act, authorizing the Secretary of the Treasury to grant the Commissioner (then Harry Anslinger) and agents of the Treasury Department's Bureau of Narcotics absolute administrative regulatory, and police powers in the enforcement of the law. The message becomes entirely clear when, having finished the short text of the Act itself, one proceeds to the sixty-odd pages of administrative and enforcement procedures established by the infamous Regulations No. 1. That regulation, not fully reproduced here, calls for a maze of affidavits, depositions, sworn statements, and constant Treasury Department police inspection in every instance that marijuana is bought, sold, used, raised, distributed, given away, and so on. Physicians who wish to purchase the one-dollar tax

stamp so that they might prescribe it for their patients are forced to report such use to the Federal Bureau of Narcotics in sworn and attested detail, revealing the name and address of the patient, the nature of his ailment, the dates and amounts prescribed, and so on. If a physician for any reason fails to do so immediately, both he and his patient are liable to imprisonment-and a heavy fine. Obviously, the details of that regulation make it far too risky for anyone to have anything to do with marijuana in any way whatsoever.

Regulations No. 1 was more than an invasion of the traditional right of privacy between patient and physician; it was a hopelessly involved set of rules that were obviously designed not merely to discourage but to prohibit the medical and popular use of marijuana. In addition to the Marihuana Tax Act and Regulations No. 1, the Bureau of Narcotics prepared a standard bill for marihuana that more than forty state legislatures enacted. This bill made possession and use of marihuana illegal per se, and so reinforced the federal act.

**U. S. TREASURY DEPARTMENT
BUREAU OF NARCOTICS
REGULATIONS No. 1
RELATING TO THE
IMPORTATION, MANUFACTURE, PRODUCTION
COMPOUNDING, SALE, DEALING IN, DISPENSING
PRESCRIBING, ADMINISTERING, AND
GIVING AWAY OF MARIHUANA UNDER THE
ACT OF AUGUST 2, 1937
PUBLIC, No. 238, 75TH CONGRESS
NARCOTIC-INTERNAL REVENUE REGULATIONS
JOINT MARIHUANA REGULATIONS MADE BY THE
COMMISSIONER OF NARCOTICS AND THE COMMISSIONER
OF INTERNAL REVENUE WITH THE APPROVAL OF THE
SECRETARY OF THE TREASURY EFFECTIVE DATE,
OCTOBER 1, 1937
LAW AND REGULATIONS RELATING TO THE IMPORTATION,
MANUFACTURE, PRODUCTION, COMPOUNDING, SALE,
DEALING IN, DISPENSING, PRESCRIBING, ADMINISTERING,
AND GIVING AWAY OF MARIHUANA**

THE LAW

(Act of Aug. 2, 1937, Public 238, 75th Congress)

Be it enacted by the Senate and House of Representatives of the United States of America in Congress assembled, that when used in this Act,

(a) The term "person" means an individual, a partnership, trust, association, company, or corporation and includes an officer or employee of a trust, association, company, or corporation, or a member or employee of a partnership, who, as such officer, employee, or member, is under a duty to perform any act in respect of which any violation of this Act occurs.

(b) The term "marihuana" means all parts of the plant Cannabis sativa L., whether growing or not; the seeds thereof; the resin extracted from any part of such plant; and every compound, manufacture, salt, derivative, mixture, or preparation of such plant, its seeds, or resin- but shall not include the mature stalks of such plant, fiber produced from such stalks, oil or cake made from the seeds of such plant, any other compound, manufacture, salt, derivative, mixture, or preparation of such mature stalks (except the resin extracted there from), fiber, oil, or cake, or the sterilized seed of such plant which is incapable of germination.

(c) The term "producer" means any person who (1) plants, cultivates, or in any way facilitates the natural growth of marihuana; or (2) harvests and transfers or makes use of marihuana.

(d) The term "Secretary" means the Secretary of the Treasury and the term "collector means collector of internal revenue.

(e) The term "transfer" or "transferred" means any type of disposition resulting in a change of possession but shall not include a transfer to a common carrier for the purpose of transporting marihuana.

SEC. 2. (a) Every person who imports, manufactures, produces, compounds, sells, deals in, dispenses, prescribes, administers, or gives away marihuana shall (1) within fifteen days after the effective date of this Act, or (2) before engaging after the expiration of such fifteen-day period in any of the above mentioned activities, and (3) thereafter, on or before July 1 of each year, pay the following special taxes respectively:

(1) Importers, manufacturers, and compounders of marihuana, $24 per year.

(2) Producers of marihuana (except those included within subdivision (4) of this subsection), $1 per year, or fraction thereof, during which they engage in such activity.

(3) Physicians, dentists, veterinary surgeons, and other practitioners who distribute, dispense, give away, administer, or prescribe marihuana to patients upon whom they in the course of their professional practice are in attendance, $1 per year or fraction thereof during which they engage in any of such activities.

(4) Any person not registered as an importer, manufacturer, producer, or compounder who obtains and uses marihuana in a laboratory for the purpose of research, instruction, or analysis, or who produces marihuana for any such purpose, $1 per year, or fraction thereof, during which he engages in such activities.

(5) Any person who is not a physician, dentist, veterinary surgeon, or other practitioner and who deals in, dispenses, or gives away marihuana, $3 per year: Provided, That any person who has registered and paid the special tax as an importer, manufacturer, compounder, or producer, as required by subdivisions (1) and (2) of this subsection, may deal in, dispense, or give away marihuana imported, manufactured, compounded, or produced by him without further payment of the tax imposed by this section.

(b) Where a tax under subdivision (1) or (5) is payable on July 1 of any year it shall be computed for one year; where any such tax is payable on any other day it shall be computed proportionately from the first day of the month in which the liability for the tax accrued to the following July 1.

(c) In the event that any person subject to a tax imposed by this section engages in any of the activities enumerated in subsection (a) of this section at more than one place, such person shall pay the tax with respect to each such place.

(d) Except as otherwise provided, whenever more than one of the activities enumerated in subsection (a) of this section is carried on by the same person at the same time, such person shall pay the tax for each such activity, according to the respective rates prescribed.

(e) Any person subject to the tax imposed by this section shall, upon payment of such tax, register his name or style and his place or places of business with the collector of the district in which such place or places of business are located.

(f) Collectors are authorized to furnish, upon written request, to any person a certified copy of the names of any or all persons who may be listed in their respective collection districts as special taxpayers under this section, upon payment of a fee of $1 for each one hundred of such names or fraction thereof upon such copy so requested.

SEC. 3. (a) No employee of any person who has paid the special tax and registered, as required by section 2 of this Act, acting within the scope of his employment, shall be required to register and pay such special tax.

(b) An officer or employee of the United States, any State, Territory, the District of Columbia, or insular possession, or political subdivision, who, in the exercise of his official duties, engages in any of the activities enumerated in section 2 of this Act, shall not be required to register or pay the special tax, but his right to this exemption shall be evidenced in such manner as the Secretary may by regulations prescribe.

SEC. 4. (a) It shall be unlawful for any person required to register and pay the special tax under the provisions of section 2 to import, manufacture, produce, compound, sell, deal in, dispense, distribute, prescribe, administer, or give away marihuana without having so registered and paid such tax.

(b) In any suit or proceeding to enforce the liability imposed by this section or section 2, if proof is made that marihuana was at any time growing upon land under the control of the defendant, such proof shall be presumptive evidence that at such time the defendant was a producer and liable under this section as well as under section 2.

SEC. 5. It shall be unlawful for any person who shall not have paid the special tax and registered, as required by section 2, to send, ship, carry, transport, or deliver any marihuana within any Territory, the District of Columbia, or any insular possession, or from any State, Territory, the District of Columbia, any insular possession of the United States, or the Canal Zone, into any other State, Territory, the District of Columbia, or insular possession of the United States: Provided, That nothing contained in this section shall apply to any common carrier engaged in transporting marihuana; or to any employee of any person who shall have registered and paid the special tax as required by section 2 while acting within the scope of his employment; or to any person who shall deliver marihuana which has been prescribed or dispensed by a physician, dentist, veterinary surgeon, or other practitioner registered under section 2, who has been employed to prescribe for the particular patient receiving such marihuana; or to any United States, State, county, municipal, District, Territorial, or insular officer or official acting within the scope of his official duties.

SEC. 6. (a) It shall be unlawful for any person, whether or not required to pay a special tax and register under section 2, to transfer marihuana, except in pursuance of a written order of the person to whom such marihuana is transferred, on a form to be issued in blank for that purpose by the Secretary.

(b) Subject to such regulations as the Secretary may prescribe, nothing contained in this section shall apply:

(1) To a transfer of marihuana to a patient by a physician, dentist, veterinary surgeon, or other practitioner registered under section 2, in the course of his professional practice only: Provided, That such physician, dentist, veterinary surgeon, or other practitioner shall keep a record of all such marihuana transferred, showing the amount transferred and the name and address of the patient to whom such marihuana is transferred, and such record shall be kept for a period of two years from the date of the transfer of such marihuana, and subject to inspection as provided in section 11.

(2) To a transfer of marihuana, made in good faith by a dealer to a consumer under and in pursuance of a written prescription issued by a physician, dentist, veterinary surgeon, or other practitioner registered under section 2: Provided, That such prescription shall be dated as of the day on which signed and shall be signed by the physician, dentist, veterinary surgeon, or other practitioner who issues the same; Provided further, That such dealer shall preserve such prescription for a period of two years from the day on which such prescription is filled so as to be readily accessible for inspection by the officers, agents, employees, and officials mentioned in section 11.

(3) To the sale, exportation, shipment, or delivery of marihuana by any person within the United States, any Territory, the District of Columbia, or any of the insular possessions of the United States, to any person in any foreign country regulating the entry of marihuana, if such sale, shipment, or delivery of marihuana is made in accordance with such regulations for importation into such foreign country as are prescribed by such foreign country, such regulations to be promulgated from time to time by the Secretary of State of the United States.

(4) To a transfer of marihuana to any officer or employee of the United States Government or of any State, Territorial, District, county, or municipal or insular government lawfully engaged in making purchases thereof for the various departments of the Army and Navy, the Public Health Service, and for Government, State, Territorial, District, county, or municipal or insular hospitals or prisons.

(S) To a transfer of any seeds of the plant Cannabis sativa L. to any person registered under section 2.

(c) The Secretary shall cause suitable forms to be prepared for the purposes before mentioned and shall cause them to be distributed to collectors for sale. The price at which such forms shall be sold by said collectors shall be fixed by the Secretary but shall not exceed 2 cents each. Whenever any collector shall sell any of such forms he shall cause the date of sale, the name and address of the proposed vendor, the name and address of the purchaser, and the amount of marihuana ordered to be plainly written or stamped thereon before delivering the same.

(d) Each such order form sold by a collector shall be prepared by him and shall include an original and two copies, any one of which shall be admissible in evidence as an original. The original and one copy shall be given by the collector to the purchaser thereof. The original shall in turn be given by the purchaser thereof to any person who shall, in pursuance thereof, transfer marihuana to him and shall be preserved by such person for a period of two years so as to be readily accessible for inspection by any officer, agent, or employee mentioned in section 11. The copy given to the purchaser by the collector shall be retained by the purchaser and preserved for a period of two years so as to be readily accessible to inspection by any officer, agent, or employee mentioned in section 11. The second copy shall be preserved in the records of the collector.

SEC. 7. (a) There shall be levied, collected, and paid upon all transfers of marihuana which are required by section 6 to be carried out in pursuance of written order forms taxes at the following rates:

(1) Upon each transfer to any person who has paid the special tax and registered under section 2 of this Act, $1 per ounce of marihuana or fraction thereof

(2) Upon each transfer to any person who has not paid the special tax and registered under section 2 of this Act, $100 per ounce of marihuana or fraction thereof.

(b) Such tax shall be paid by the transferee at the time of securing each order form and shall be in addition to the price of such form. Such transferee shall be liable for the tax imposed by this section but in the event that the transfer is made in violation of section 6 without an order form and without payment of the transfer tax imposed by this section, the transferor shall also be liable for such tax.

(c) Payment of the tax herein provided shall be represented by appropriate stamps to be provided by the Secretary and said stamps shall be affixed by the collector or his representative to the original order form.

(d) All provisions of law relating to the engraving, issuance, sale, accountability, cancellation, and destruction of tax-paid stamps provided for in the internal-revenue laws shall, insofar as applicable and not inconsistent with this Act, be extended and made to apply to stamps provided for in this section.

(e) All provisions of law (including penalties) applicable in respect of the taxes imposed by the Act of December 17, 1914 (38 Stat. 785; U. S. C., 1934 ed., title 26, secs. 1040-- 1061, 1383-1391), as amended, shall, insofar as not inconsistent with this Act, be applicable in respect of the taxes imposed by this Act.

SEC. 8. (a) It shall be unlawful for any person who is a transferee required to pay the transfer tax imposed by section 7 to acquire or otherwise obtain any marihuana without having paid such tax; and proof that any person shall have had in his possession any marihuana and shall have failed, after reasonable notice and demand by the collector, to produce the order form required by section 6 to be retained by him, shall be presumptive evidence of guilt under this section and of liability for the tax imposed by section 7.

(b) No liability shall be imposed by virtue of this section upon any duly authorized officer of the Treasury Department engaged in the enforcement of this Act or upon any duly authorized officer of any State, or Territory, or of any political subdivision thereof, or the District of Columbia, or of any insular possession of the United States, who shall be engaged in the enforcement of any law or municipal ordinance dealing with the production, sale, prescribing, dispensing, dealing in, or distributing of marihuana.

SEC. 9. (a) Any marihuana which has been imported, manufactured, compounded, transferred, or produced in violation of any of the provisions of this Act shall be subject to seizure and forfeiture and, except as inconsistent with the provisions of this Act, all the provisions of internal-revenue laws relating to searches, seizures, and forfeitures are extended to include marihuana.

(b) Any marihuana which may be seized by the United States Government from any person or persons charged with any violation of this Act shall upon conviction of the person or persons from whom seized be confiscated by and forfeited to the United States.

(c) Any marihuana seized or coming into the possession of the United States in the enforcement of this Act, the owner or owners of which are unknown, shall be confiscated by and forfeited to the United States.

(d) The Secretary is hereby directed to destroy any marihuana confiscated by and forfeited to the United States under this section or to deliver such marihuana to any department, bureau, or other agency of the United States Government, upon proper application therefore under such regulations as may be prescribed by the Secretary.

SEC. 10. (a) Every person liable to any tax imposed by this act shall keep such books and records, render under oath such statements, make such returns, and comply with such rules and regulations as the Secretary may from time to time prescribe.

(b) Any person who shall be registered under the provisions of section 2 in any internal- revenue district shall, whenever required so to do by the collector of the district, render to the collector a true and correct statement or return, verified by affidavits, setting forth the quantity of marihuana received or harvested by him during such period immediately preceding the demand of the collector, not exceeding three months, as the said collector may fix and determine. If such person is not solely a producer, he shall set forth in such statement or return the names of the persons from which said marihuana was received, the quantity in each instance received from such persons, and the date when received.

SEC. 11. The order forms and copies thereof and the prescriptions and records required to be preserved under the provisions of section 6, and the statements or returns filed in the office of the collector of the district under the provisions of section 10 (b) shall be open to inspection by officers, agents, and employees of the Treasury Department duly authorized for that purpose, and such officers of any State, or Territory, or of any political subdivision thereof, or the District of Columbia, or of any insular possession of the United States as shall be charged with the enforcement of any law or municipal ordinance regulating the production, sale, prescribing, dispensing, dealing in, or distributing of marihuana. Each collector shall be authorized to furnish, upon written request, copies of any of the said statements or returns filed in his office to any of such officials of any State or Territory, or political subdivision thereof, or the District of Columbia, or any insular possession of the United States as shall be entitled to inspect the said statements or returns filed in the office of the said collector, upon the payment of a fee of $1 for each 100 words or fraction thereof in the copy or copies so requested.

SEC. 12. Any person who is convicted of a violation of any provision of this Act shall be fined not more than $2,000 or imprisoned not more than five years, or both, in the discretion of the court.

SEC. 13. It shall not be necessary to negative any exemptions set forth in this Act in any complaint, information, indictment, or other writ or proceeding laid or brought under this Act and the burden of proof of any such exemption shall be upon the defendant. In the absence of the production of evidence by the defendant that he has complied with the provisions of section 6 relating to order forms, he shall be presumed not to have complied with such provisions of such sections, as the case may be.

SEC. 14. The Secretary is authorized to make, prescribe, and publish all necessary rules and regulations for carrying out the provisions of this Act and to confer or impose any of the rights, privileges, powers, and duties conferred or imposed upon him by this Act upon such officers or employees of the Treasury Department as he shall designate or appoint.

SEC. 15. The provisions of this Act shall apply to the several States, the District of Columbia, the Territory of Alaska, the Territory of Hawaii, and the insular possessions of the United States, except the Philippine Islands. In Puerto Rico the administration of this Act, the collection of the special taxes and transfer taxes, and the issuance of the order forms provided for in section 6 shall be performed by the appropriate internal revenue officers of that government, and all revenues collected under this Act in Puerto Rico shall accrue intact to the general government thereof. The President is hereby authorized and directed to issue such Executive orders as will carry into effect in the Virgin Islands the intent and purpose of this Act by providing for the registration with appropriate officers and the imposition of the special and transfer taxes upon all persons in the Virgin Islands who import, manufacture, produce, compound, sell, deal in, dispense, prescribe, administer, or give away marihuana.

SEC. 16. If any provision of this Act or the application thereof to any person or circumstances is held invalid, the remainder of the Act and the application of such provision to other persons or circumstances shall not be affected thereby.

SEC. 17. This Act shall take effect on the first day of the second month during which it is enacted.

SEC. 18. This Act may be cited as the "Marihuana Tax Act of 1937." (T. D. 28)

Order of the Secretary of the Treasury Relating to the Enforcement of the Marihuana Tax Act of 1937

September 1, 1937

Section 14 of the Marihuana Tax Act of 1937 (act of Congress approved August 2, 1937, Public, No. 238), provides as follows:

The Secretary is authorized to make, prescribe, and publish all necessary rules and regulations for carrying out the provisions of this Act and to confer or impose any of the rights, privileges, powers, and duties conferred or imposed upon him by this Act upon such officers or employees of the Treasury Department as he shall designate or appoint.

In pursuance of the authority thus conferred upon the Secretary of the Treasury, it is hereby ordered:

1. Rights, Privileges, Powers, and Duties Conferred and imposed Upon the Commissioner of Narcotics

1. There are hereby conferred and imposed upon the Commissioner of Narcotics, subject to the general supervision and direction of the Secretary of the Treasury, all the rights, privileges, powers, and duties conferred or imposed upon said Secretary by the Marihuana Tax Act of 1937, so far as such rights privileges, powers, and duties relate to:

(a) Prescribing regulations, with the approval of the Secretary, as to the manner in which the right of public officers to exemption from registration and payment of special tax may be evidenced, in accordance with section 3 (b) of the act.

(b) Prescribing the form of written order required by section 6 (a) of the act, said form to be prepared and issued in blank by the Commissioner of Internal Revenue as hereinafter provided.

(c) Prescribing regulations, with the approval of the Secretary, giving effect to the exceptions, specified in subsection (b), from the operation of subsection (a) of section 6 of the act.

(d) The destruction of marihuana confiscated by and forfeited to the United States, or delivery of such marihuana to any department, bureau, or other agency of the United States Government, and prescribing regulations, with the approval of the Secretary, governing the manner of application for, and delivery of such marihuana.

(e) Prescribing rules and regulations, with the approval of the Secretary, as to books and records to be kept, and statements and information returns to be rendered under oath, as required by section 10 (a) of the act.

(f) The compromise of any criminal liability (except as relates to delinquency in registration and delinquency in payment of tax) arising under the act, in accordance with section 3229 of the Revised Statutes of the United States (U. S. Code (1934 ed.) title 26, sec. 1661), and the recommendation for assessment of civil liability for internal- revenue taxes and ad valorem penalties under the act.

II. Rights, Privileges, Powers, and Duties Conferred and Imposed upon the Commissioner of Internal Revenue.

1. There are hereby conferred and imposed upon the Commissioner of Internal Revenue, subject to the general supervision and direction of the Secretary of the Treasury, the rights, privileges,

powers, and duties conferred or imposed upon said Secretary of the Marihuana Tax Act of 1937, not otherwise assigned herein, so far as such rights, privileges, powers, and duties relate to

(a) Preparation and issuance in blank to collectors of internal revenue of the written orders, in the form prescribed by the Commissioner of Narcotics, required by section 6 (a) of the act. The price of the order form, as sold by the collector under section 6 (c) of the act shall be two cents for the original and one copy.

(b) Providing appropriate stamps to represent payment of transfer tax levied by section 7, and prescribing and providing appropriate stamps for issuance of special tax payers registering under section 2 of the act.

(c) The compromise of any civil liability involving delinquency in registration, delinquency in payment of tax, and ad valorem penalties, and of any criminal liability incurred through delinquency in registration and delinquency in payment of tax, in connection with the act and in accordance with Section 3229 of the Revised Statutes of the United States (U. S. Code (1934 ed.), title 26, sec. 1661)- the determination of liability for and the assessment and collection of special and transfer taxes imposed by the act; the determination of liability for and the assessment and collection of the ad valorem penalties imposed by Section 3176 of the Revised Statutes, as modified by Section 406 of the Revenue Act of 1935 (U. S. Code (1934 ed.) title 26, secs. 1512-1525), for delinquency in registration; and the determination of liability for and the assertion of the specific penalty imposed by the act, for delinquency in registration and payment of tax.

GENERAL PROVISIONS

The investigation and the detection, and presentation to prosecuting officers of evidence, of violations of the Marihuana Tax Act of 1937, shall be the duty of the Commissioner of Narcotics and the assistants, agents, inspectors, or employees under his direction. Except as specifically inconsistent with the terms of said act and of this order, the Commissioner of Narcotics and the Commissioner of Internal Revenue and the assistants, agents, inspectors, or employees of the Bureau of Narcotics and the Bureau of Internal Revenue, respectively, shall have the same powers and duties in safeguarding the revenue thereunder as they now have with respect to the enforcement of, and collection of the revenue under, the act of December 17, 1914, as amended (U. S. Code (1934 ed.), title 26, sec. 1049).

In any case where a general offer is made in compromise of civil and criminal liability ordinarily compromisable hereunder by the Commissioner of Internal Revenue and of criminal liability ordinarily compromisable hereunder by the Commissioner of Narcotics, the case may be jointly compromisable by those officers, in accordance with Section 3229 of the Revised Statutes of the United States (U. S. Code (1934 ed.), title 26, sec. 1661).

Power is hereby conferred upon the Commissioner of Narcotics to prescribe such regulations as he may deem necessary for the execution of the functions imposed upon him or upon the officers or employees of the Bureau of Narcotics, but all regulations and changes in regulations shall be subject to the approval of the Secretary of the Treasury.

The Commissioner of Internal Revenue and the Commissioner of Narcotics may, if they are of the opinion that the good of the service will be promoted thereby, prescribe regulations relating to internal revenue taxes where no violation of the Marihuana Tax Act of 1937 is involved, jointly, subject to the approval of the Secretary of the Treasury.

The right to amend or supplement this order or any provision thereof from time to time, or to revoke this order or any provision thereof at any time, is hereby reserved.

The effective date of this order shall be October 1, 1937, which is the effective date of the Marihuana Tax Act of 1937.

STEPHEN B. GIBBONS,
Acting Secretary of the Treasury.

REGULATIONS

Introductory

The Marihuana Tax Act of 1937, imposes special (occupational) taxes upon persons engaging in activities involving articles or material within the definition of "marihuana" contained in the act, and also taxes the transfer of such articles or material.

These regulations deal with details as to tax computation, procedure, the forms of records and returns, and similar matters. These matters in some degree are controlled by certain sections of the United States Revised Statutes and other statutes of general application. Provisions of these statutes, as well as of the Marihuana Tax Act of 1937 are quoted, in whole or in part, as the immediate or general basis for the regulatory provisions set forth. The quoted provisions are from the Marihuana Tax Act of 1937 unless otherwise indicated.

Provisions of the statutes upon which the various articles of the regulations are based generally have not been repeated in the articles.

144

Therefore, the statutory excerpts preceding the several articles should be examined to obtain complete information.

Chapter I

Laws Applicable

SEC. 7 (e) All provisions of law (including penalties) applicable in respect of the taxes imposed by the Act of December 17, 1914 (38 Stat. 785; U. S. C., 1934 ed., title 26, secs. 1040- 1061, 1383-1391), as amended, shall, insofar as not inconsistent with this Act, be applicable in respect of the taxes imposed by this Act.

ART. 1. Statutes applicable. All general provisions of the internal revenue laws, not inconsistent with the Marihuana Tax Act, are applicable in the enforcement of the latter.

Chapter II

Definitions

SEC. 1. That when used in this Act:

(a) The term "person" means an individual, a partnership, trust, association, company, or corporation and includes an officer or employee of a trust, association, company, or corporation, or a member or employee of a partnership, who as such officer, employee, or member is under a duty to perform . any act in respect of which any violation of this Act occurs.

(b) The term "marihuana" means all parts of the plant Cannabis sativa L., whether growing or not; the seeds thereof; the resin extracted from any part of such plant; and every compound, manufacture, salt, derivative, mixture, or preparation of such plant, its seeds, or resins; but shall not include the mature stalks of such plant, fiber produced from such stalks, oil or cake made from the seeds of such plant, any other compound, manufacture, salt, derivative, mixture, or preparation of such mature stalks (except the resin extracted therefrom), fiber, oil, or cake, or the sterilized seed of such plant which is incapable of germination.

(c) The term "producer" means any person who (1) plants, cultivates, or in any way facilitates the natural growth of marihuana; or (2) harvests and transfers or makes use of marihuana.

(d) The term "Secretary" means the Secretary of the Treasury and the term "collector" means collector of internal revenue.

(e) The term "transfer" or "transferred" means any type of disposition resulting in a change of possession but shall not include a transfer to a common carrier for the purpose of transporting marihuana

ART. 2. As used in these regulations:

(a) The term "act" or "this act" shall mean the Marihuana Tax Act of 1937, unless otherwise indicated.

(b) The term "United States" shall include the several States, the District of Columbia, the Territory of Alaska, the Territory of Hawaii, and the insular possessions of the United States except Puerto Rico and the Virgin Islands. It does not include the Canal Zone or the Philippine Islands.

(c) The terms "manufacturer" and "compounder" shall include any person who subjects marihuana to any process of separation, extraction, mixing, compounding, or other manufacturing operation. They shall not include one who merely gathers and destroys the plant, one who merely threshes out the seeds on the premises where produced, or one who in the conduct of a legitimate business merely subjects seeds to a cleaning process.

(d) The term "producer" means any person who induces in any way the growth of marihuana, and any person who harvests it, either in a cultivated or wild state, from his own or any other land, and transfers or makes use of it, including one who subjects the marihuana which he harvests to any processes rendering him liable also as a manufacturer or compounder. Generally all persons are included who gather marihuana for any purpose other than to destroy it. The term does not include one who merely plows under or otherwise destroys marihuana with or without harvesting. It does not include one who grows marihuana for use in his own laboratory for the purpose of research, instruction, or analysis and who does not use it for any other purpose or transfer it.

(e) The term "special tax" is used to include any of the taxes, pertaining to the several occupations or activities covered by the act, imposed upon persons who import, manufacture, produce, compound, sell, deal in, dispense, prescribe, administer, or give away marihuana.

(f) The term "person" occurring in these regulations is used to include individual, partnership, trust, association, company, or corporation; also a hospital, college of pharmacy, medical or dental clinic, sanatorium, or other institution or entity.

(g) Words importing the singular may include the plural; words importing the masculine gender may be applied to the feminine or the neuter.

The definitions contained herein shall not be deemed exclusive.

Congressman Ron Paul

United States House of Representatives Bill H.R. 3037

Industrial Hemp Farming Act of 2005

Title: To amend the Controlled Substances Act
to exclude industrial hemp from the definition of
marihuana, and for other purposes.
Sponsor: Rep Paul, Ron [TX-14-photo right]
(introduced 6/22/2005)
Cosponsors (11)
Latest Major Action: 7/1/2005
Referred to House subcommittee.
Status: Referred to the Subcommittee on Health.
SUMMARY AS OF: 6/22/2005--Introduced.

Industrial Hemp Farming Act of 2005 - Amends the Controlled Substances Act to exclude industrial hemp from the definition of "marihuana." Defines "industrial hemp" to mean the plant Cannabis sativa L. and any part of such plant with a delta-nine tetrahydrocannabinol concentration that does not exceed .3 percent on a dry weight basis. Grants a state regulating the growing and processing of industrial hemp exclusive authority, in any criminal or civil action or administrative proceeding, to determine whether any such plant meets that concentration limit.

MAJOR ACTIONS:***NONE***
ALL ACTIONS:
6/22/2005:
Introductory remarks on measure. (CR E1313-1314)
6/22/2005:
Referred to the Committee on Energy and Commerce, and in addition to the Committee on the Judiciary, for a period to be subsequently determined by the Speaker, in each case for consideration of such provisions as fall within the jurisdiction of the committee concerned.
6/22/2005: Referred to House Energy and Commerce
7/1/2005: Referred to the Subcommittee on Health.
6/22/2005: Referred to House Judiciary
TITLE(S): (italics indicate a title for a portion of a bill) ***NONE***

COSPONSORS(11),
ALPHABETICAL [followed by Cosponsors withdrawn]:
Industrial Hemp Farming Act of 2005 (Introduced in House)
HR 3037 IH
109th CONGRESS
1st Session
H. R. 3037
To amend the Controlled Substances Act to exclude industrial hemp from the definition of marihuana, and for other purposes.
IN THE HOUSE OF REPRESENTATIVES
June 22, 2005
Mr. PAUL (for himself, Mr. FARR, Mr. MCDERMOTT, Mr. STARK, and Mr. GRIJALVA) introduced the following bill; which was referred to the Committee on Energy and Commerce, and in addition to the Committee on the Judiciary, for a period to be subsequently determined by the Speaker, in each case for consideration of such provisions as fall within the jurisdiction of the committee concerned

A BILL

To amend the Controlled Substances Act to exclude industrial hemp from the definition of marihuana, and for other purposes.

Be it enacted by the Senate and House of Representatives of the United States of America in Congress assembled,

SECTION 1. SHORT TITLE.

This Act may be cited as the `Industrial Hemp Farming Act of 2005'.

SEC. 2. EXCLUSION OF INDUSTRIAL HEMP FROM DEFINITION OF MARIHUANA.

Paragraph (16) of section 102 of the Controlled Substances Act (21 U.S.C. 802(16)) is amended--

(1) by striking `(16)' at the beginning and inserting `(16)(A)'; and

(2) by adding at the end the following new subparagraph:

(B) The term `marihuana' does not include industrial hemp. As used in the preceding sentence, the term `industrial hemp' means the plant Cannabis sativa L. and any part of such plant, whether growing or not, with a delta-9 tetrahydrocannabinol concentration that does not exceed 0.3 percent on a dry weight basis.'.

SEC. 3. INDUSTRIAL HEMP DETERMINATION
TO BE MADE BY STATES.

Section 201 of the Controlled Substances Act (21 U.S.C. 811) is amended by adding at the end the following new subsection:

`(i) Industrial Hemp Determination to Be Made by States- In any criminal action, civil action, or administrative proceeding, a State regulating the growing and processing of industrial hemp under State law shall have exclusive authority to determine whether any such plant meets the concentration limitation set forth in subparagraph (B) of paragraph (16) of section 102 and such determination shall be conclusive and binding.'.

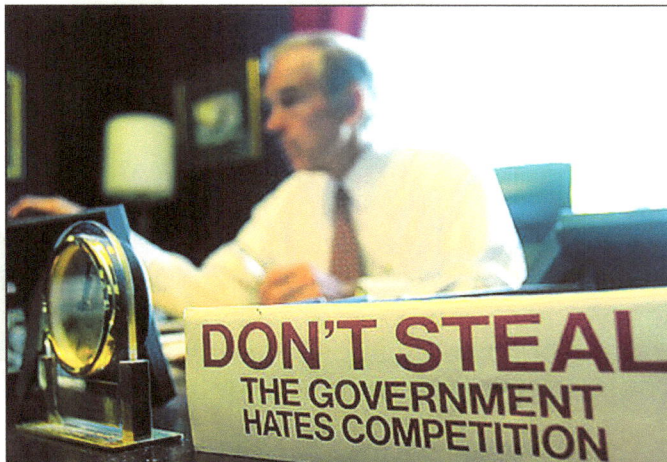

Industrial Hemp Farming Act Sponsor
Congressman Ron Paul's Desk sign:

DON'T STEAL

THE GOVERNMENT HATES COMPETITION

WWW.HOUSE.GOV/PAUL

Industrial Hemp Farming Act of 2007

HR 1009 IH

110TH CONGRESS

1ST SESSION

H. R. 1009

To amend the Controlled Substances Act to exclude industrial hemp from the definition of marihuana, and for other purposes.

IN THE HOUSE OF REPRESENTATIVES
February 13, 2007

Mr. PAUL (for himself, Ms. BALDWIN, Mr. FRANK of Massachusetts, Mr. GRIJALVA, Mr. HINCHEY, Mr. KUCINICH, Mr. MCDERMOTT, Mr. GEORGE MILLER of California, Mr. STARK, and Ms. WOOLSEY) introduced the following bill; which was referred to the Committee on Energy and Commerce, and in addition to the Committee on the Judiciary, for a period to be subsequently determined by the Speaker, in each case for consideration of such provisions as fall within the jurisdiction of the committee concerned

A BILL

To amend the Controlled Substances Act to exclude industrial hemp from the definition of marihuana, and for other purposes.

Be it enacted by the Senate and House of Representatives of the United States of America in Congress assembled,

SECTION 1. SHORT TITLE.

This Act may be cited as the `Industrial Hemp Farming Act of 2007'.

SEC. 2. EXCLUSION OF INDUSTRIAL HEMP
FROM DEFINITION OF MARIHUANA.

Paragraph (16) of section 102 of the Controlled Substances Act (21 U.S.C. 802(16)) is amended--

(1) by striking `(16)' at the beginning and inserting `(16)(A)'; and

(2) by adding at the end the following new subparagraph:

`(B) The term `marihuana' does not include industrial hemp. As used in the preceding sentence, the term `industrial hemp' means the plant Cannabis sativa L. and any part of such plant, whether growing or not, with a delta-9 tetrahydrocannabinol concentration that does not exceed 0.3 percent on a dry weight basis.'

SEC. 3. INDUSTRIAL HEMP DETERMINATION
TO BE MADE BY STATES.

Section 201 of the Controlled Substances Act (21 U.S.C. 811) is amended by adding at the end the following new subsection:

`(i) Industrial Hemp Determination To Be Made by States- In any criminal action, civil action, or administrative proceeding, a State regulating the growing and processing of industrial hemp under State law shall have exclusive authority to determine whether any such plant meets the concentration limitation set forth in subparagraph (B) of paragraph (16) of section 102 and such determination shall be conclusive and binding.'.

Not Passed—Lost In Committee Since 4/20/07

13. 50 THINGS YOU CAN DO
TO FIGHT GLOBAL WARMING

1. Learn about global warming and see Al Gore's film on global warming, "An Inconvenient Truth." It is also in book form. Buy a copy for your kids, parents, friends, etc.

2. Walk, bike, carpool or use public transportation whenever possible. Active pursuit of personal ways out of the 'great car economy.'

3. Recycle and buy minimally packed goods as much as possible.

4. Support a sixty mile per hour maximum speed limit with teeth. Support strict mileage standards.

5. Support green power initiatives of power companies to supply non-polluting energy.

6. Adjust your thermostat. Expand your comfort range by a few degrees.

7. Wash clothes in cold or warm water, not hot.

8. Install low-flow shower heads to use less water. Shower with a friend.

9. Run the dishwasher only when full and don't use heat to dry dishes.

10. Replace standard light bulbs with compact fluorescent bulbs.

11. Insulate the roof, the hot water tank, and the walls.

12. Plug air leaks in windows and doors to increase energy efficiency by weather-stripping and double-glazing.

13. Replace old appliances with energy-efficient models.

14. Recycling of paper and avoidance of excess packaging and disposable products. Landfill products generate the greenhouse gas methane.

15. Favoring organically farmed produce over intensive-farming products.

16. Favoring vegetarian produce over meat. Cows produce huge quantities of methane gas and one pound of meat requires more than 5,000 gallons of water to produce.

17. Exercise of consumer discretion concerning the products of companies whose activities add to the greenhouse threat.

18. Use of the power of the pen in exerting pressure for anti-greenhouse changes in society.

19. Buy hemp products, hemp food, and hemp fuel where possible. Support the growing of hemp world-wide on an emergency basis.

20. Be a future hero by not reproducing yourself. Support family planning for everyone in the world. The population of the earth increases by 260,000 people each day.

21. Join the fight to save farmland in your area. Our survival may depend on hemp farmland to reverse global warming.

22. Lobby for reforestation and to stop deforestation with hemp.

23. Add to this list whatever you can to increase our chances for survival.

24. Share these simple steps with friends and family and increase awareness!

25. Get these survival steps taught in your local school. Kids need to know the truth about how humans have impacted the earth in just a few generations.

26. Come up with 25 more things we can do to stop global warming.

14. THE U.S.A. HEMP MUSEUM CURATOR'S ROOM
Richard M. Davis
FREEDOM FIGHTER, FEB. '95 HIGH TIMES MAGAZINE

I am Richard M. Davis, founder and curator of the FIRST Virtual Traveling Hemp Museum. The USA Hemp Museum, both the Traveling Museum and the resource website www.hempmuseum.org teach the benefits of Cannabis/Hemp, as a medicine, as an industrial resource and for the private use by adults. There is also a private museum in my home brimming with exhibits in Los Angeles, California.

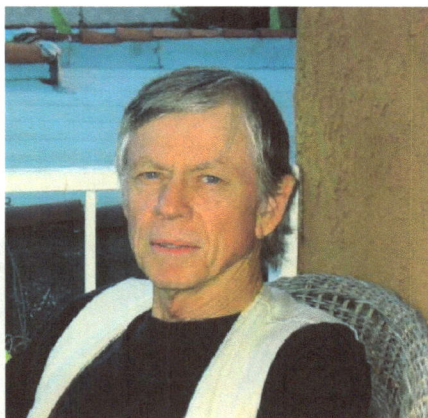

I was born in Arizona, graduated high school in Willcox, Arizona, joined the Air Force from Arizona, and was arrested for the first time in Arizona at the age of fifty-five. I reside in Los Angeles, California.

I hold a Masters Degree in Biology from California State University at Los Angeles, and attended the School of Public Health at UCLA for four years under a US Public Health Service Fellowship.

I have lived surrounded by hemp for forty of my 66 years (born 1940), the vast majority of my life. My introduction to hemp came, like that of many in my generation, from hemp smoke. I was refused the request to study the effects of pot at the School of Public Health at U.C.L.A., in 1972, where I was working on a doctorate in Public Health. I studied zoology (B.S.) and biology (M.A.).

I took a leave of absence and never went back. After a dozen years in the mountains on a small farm, I ran for Congress in a democratic primary as an admitted pot grower. Several years later when the U.S. Army invaded that area in operation Greensweep, I found out about hemp from Jack Herer's book, The Emperor Wears No Clothes, and started the Hemp Museum, with help of course.

The USA Hemp Museum
Had Been doing this a long time
And it's still true –

Hemp Is A Solution To Global Warming!

Mendocino Mobile Marijuana Museum

AT THE CALIFORNIA CAPITOL BUILDING IN SACRAMENTO.
THE FIRST MUSEUM WAS IN A HONDA WAGON,
AND THE PLANT ON TOP
SEEMED AS LARGE AS THE WHOLE CAR.

We were first the Mendocino Mobile Marijuana Museum, a knock at the government that wants to call hemp the marijuana word in law, when it does not apply. With two card tables and one hemp shirt, we passed out literature to hundreds of Capitol employees and elected officials on the benefits of legal hemp and hemp medicines.

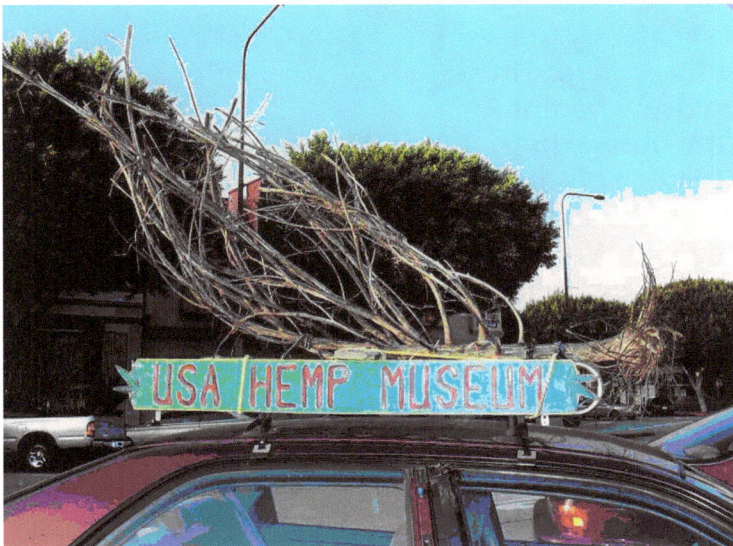

**Hempmobile
with hemp branches
proudly displayed
on top at the
2006 Los Angeles
Million Marijuana
March
hosted by the late
Sister
Somayah
Kambui**

THE USA HEMP MUSEUM TRUCK

Notice the medical hemp plant on top of the camper.

The truck brought an added dimension of high visibility and lots of bumper sticker space.

When the virtual museum started, much of the site had to do with my trial in Arizona as that was what was happening at the time. James Dawson of Florida was our dedicated webmaster, who teamed with Brenda Kershenbaum here in Los Angeles to do the initial web work. Later the gallery came into being with some 60 pictures, taken by photographer Bill Bridges, with the assistance of Tim Perkins. The plan was to reconstruct the museum into various rooms where the focus will be on some aspect of the hemp plant or the consequences of prohibition. The virtual museum now contains 1700 picture files (some repetition) and 18 virtual rooms, dealing with hemp in history, agriculture, textiles, plastics, medicine, rope, etc.

The Alterna Hemp Shampoo ads were plastered all over the Los Angeles Area for months and caused quite a stir. They got heat for displaying the most famous leaf in history. George Washington grew hemp!

What the hemp museum means in real terms is a look at the hemp history of the world, the issues of hemp prohibition, some people involved with hemp, and of course the future of hemp. We feel the world with hemp will be a better place. Our friends will be out of jail, issues will find new solutions, patients will improve, and the war will end when hemp is free of its prohibition.

WE MUST ALL DO OUR SHARE...

NO MATTER THE CONSEQUENCES,

TO <u>END</u> THIS TERRIBLE WAR ON AMERICA.

Ode For An Agricultural Celebration

By

William Cullen Bryant

Far back in the ages,
The plough with wreaths was crowned;
The hands of kings and sages
Entwined the chaplet round;
Till men of spoil disdained the toil
By which the world was nourished,
And dews of blood enriched the soil
Where green their laurels flourished:
---Now the world her fault repairs—
The guilt that stains her story;
And weeps her crimes amid the cares
That formed her earliest glory.

The proud throne shall crumble,
The diadem shall wane,
The tribes of earth shall humble
The pride of those who reign;
And War shall lay his pomp away; ---
The fame that heroes cherish,
The glory earned in deadly fray,
Shall fade, decay, and perish.
Honor waits, o'er all the Earth,
Through endless generations,
The art that calls her harvests forth,
And feeds the expectant nations.

Appendix

Industrial Hemp in the United States: Status   159

Waste to Energy (WTE) & Biomass in California   161

Reports and Papers on Biomass & Waste to Energy   162

Vermont Legislative Research Shop -   163

Hemp Oil Fuels & How To Make Them by A. Das   176

The Yearbook of The US Department Of Agriculture, 1913   180

Hemp-knowledgements   227

The Research Application   228

Hemp Products Flowchart

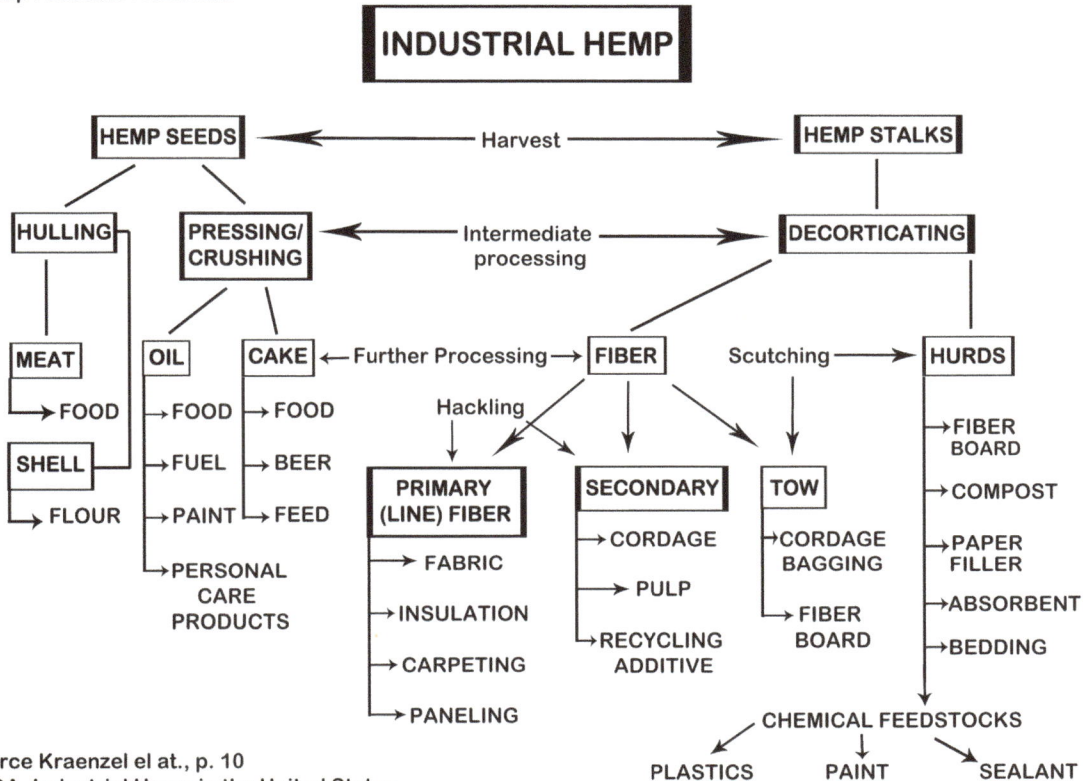

INDUSTRIAL HEMP

HEMP SEEDS ←——— Harvest ———→ HEMP STALKS

HULLING   PRESSING/CRUSHING ←—— Intermediate processing ——→ DECORTICATING

MEAT   OIL   CAKE ←— Further Processing —→ FIBER   Scutching ———→ HURDS

→FOOD   →FOOD   →FOOD   Hackling

SHELL   →FUEL   →BEER

→ FLOUR   →PAINT   →FEED

→PERSONAL CARE PRODUCTS

PRIMARY (LINE) FIBER   SECONDARY   TOW

→ FABRIC   →CORDAGE   →CORDAGE BAGGING

→INSULATION   →PULP   →FIBER BOARD

→CARPETING   →RECYCLING ADDITIVE

→PANELING

→FIBER BOARD

→COMPOST

→PAPER FILLER

→ABSORBENT

→BEDDING

CHEMICAL FEEDSTOCKS

PLASTICS   PAINT   SEALANT

Source Kraenzel el at., p. 10
USDA- Industrial Hemp in the United States

Source: US Department of Agriculture
www.ers.usda.gov/publications/ages001E/ages001E.pdf

The USDA's Industrial Hemp flow chart above
does not reveal the whole picture.

Hurds left out fuel or fuel alcohols.

Paper filler includes all types of paper
i.e. writing paper, wall paper, industrial grade packaging materials.

Cordage is a wide range of strong hemp ropes and twines.

Industrial Hemp can replace many toxic materials we use for
food, clothing, fuel, building materials, 50,000+ products.

Figure 3.
A typical breakdown of the green and dry-plant components of hemp grown for fiber

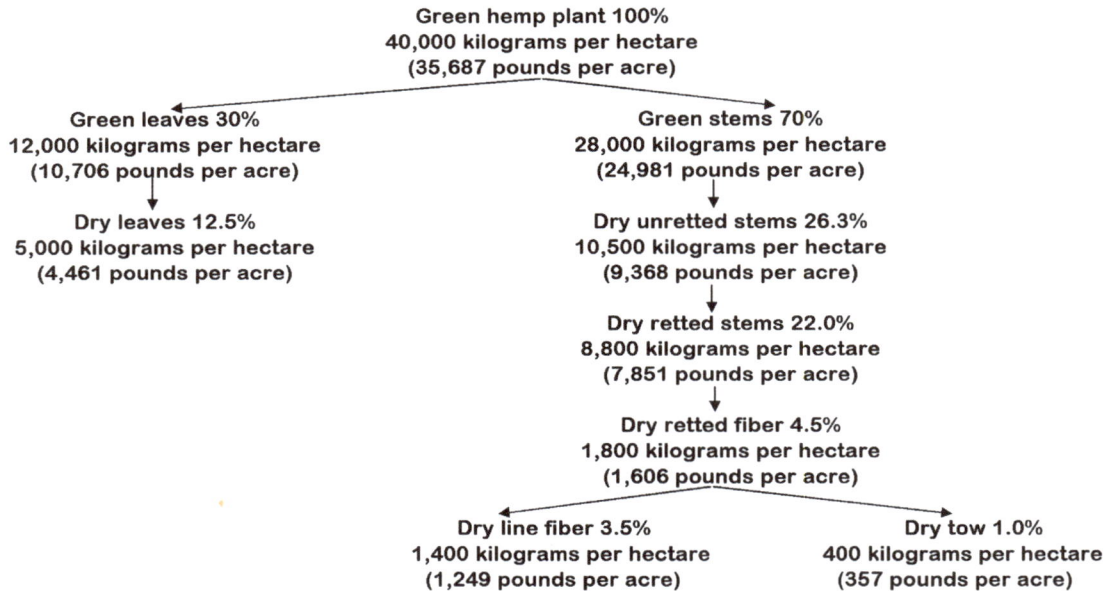

Green hemp plant 100%
40,000 kilograms per hectare
(35,687 pounds per acre)

Green leaves 30%
12,000 kilograms per hectare
(10,706 pounds per acre)

Green stems 70%
28,000 kilograms per hectare
(24,981 pounds per acre)

Dry leaves 12.5%
5,000 kilograms per hectare
(4,461 pounds per acre)

Dry unretted stems 26.3%
10,500 kilograms per hectare
(9,368 pounds per acre)

Dry retted stems 22.0%
8,800 kilograms per hectare
(7,851 pounds per acre)

Dry retted fiber 4.5%
1,800 kilograms per hectare
(1,606 pounds per acre)

Dry line fiber 3.5%
1,400 kilograms per hectare
(1,249 pounds per acre)

Dry tow 1.0%
400 kilograms per hectare
(357 pounds per acre)

Note: Although these stem and fiber yields are from 1970,
they illustrate how bast fibers are only a small portion of total crop yields.
Source: Dempsey, p. 82. USDA

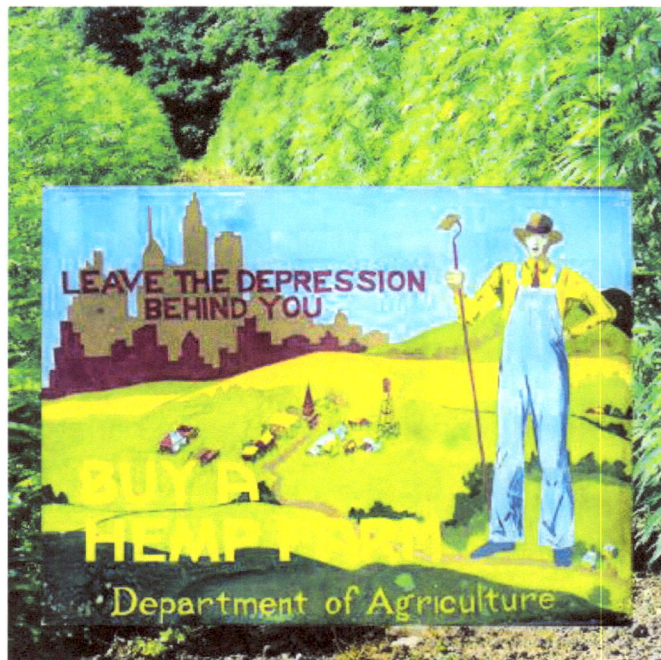

Waste to Energy (WTE) & Biomass in California

CALIFORNIA ENERGY COMMISSION
WWW.ENERGY.CA.GOV / DEVELOPMENT / BIOMASS / #BIOMASS

Californians create nearly 2,900 pounds of household garbage and industrial waste each and every second; a total of 45 million tons of waste per year (according to the California Integrated Waste Management Board)! Until recently, the only place to put that trash was in local landfills. Today, however, waste and its by-products are being recycled into more useful products. Some waste materials can also be used as a fuel in power plants to create electricity or other forms of energy.

These power plants are defined by the type of fuel source they use: biomass, digester gas, industrial waste, landfill gas, and municipal solid waste. All together there are 90 waste-to-energy plants in California with a total installed capacity of 971 megawatts. These plants produced 5,848 million kilowatt-hours of electricity in 1996, about 2.3 percent of the state's total electricity production.

BIOMASS

**Digester Gas
(Anaerobic Digestion)**

Industrial Waste

Landfill Gas

Municipal Solid Waste

Waste Tire Conferences

Reports and Papers on Biomass
& Waste to Energy

Energy Commission Activities Related to Item 11 of Governor's Executive Order D-5-99, of March 25, 1999 (Phasing out the use of MTBE in Gasoline) on the potential for a California Waste-Based or Other Biomass Ethanol Development

Energy Technology Status Report - Chapter 8: Biomass. May 1999. File is Acrobat PDF listed in table of contents.

The Growing Importance of Biomass. Article by William J. Keese, Chairman, California Energy Commission. California Biodiversity News (California Diodiversity Council), Summer 1998 - Volume 5, Number 4.

New Hope for the Tahoe Basin. Article by Rob Schlichting, California Energy Commission; and Karen Terrill and Ross Henly, Department of Forestry and Fire Protection. California Biodiversity News (California Diodiversity Council), Summer 1998 - Volume 5, Number 4.

The
UNIVERSITY
of VERMONT

Vermont Legislative Research Shop Viability of Industrial Hemp

Industrial hemp is derived from the Cannabis sativa plant, the same plant that marijuana is derived from. The two plants differ in that marijuana comes from the leaves and flowers. Industrial hemp is grown for use of the stalk and seeds. They also differ in levels of tetrahydrocannibinol, the chemical that is responsible for its psychoactive properties. Marijuana plants contain levels of 3-15% THC and plants grown for industrial hemp contain less than 1% of THC (Frohling and Staton 1997). Research has consistently shown that the low THC level in hemp plants is not capable of producing the psychoactive effects that marijuana plants do (Hawaii House of Representatives). Three main raw materials may be produced from industrial hemp plants: bast fiber, hurds and seeds. Industrial hemp is a very versatile product; it produces textiles, rope, cellulose plastics, resins, particleboard, paper products and oil. It is one of the strongest natural fibers, is a high quality absorbent and is recyclable. All hemp-based products are biodegradable. Hemp seeds contain 20-25% protein and are high in calcium, magnesium, phosphorous, potassium and vitamin A. Hemp seed oil is high in essential fatty acids (EFAs) that help lower cholesterol levels. It is used in various foods and to make non-dairy products. The oil is also used for cosmetics, paints and varnishes, inks, and when combined with 15% methanol a substitute for diesel fuel is produced that burns 70% cleaner than petroleum diesel (Hawaii House of Representatives). See Appendix A for a report from the National Conference of State Legislatures on the utility of hemp.

The pursuit and success rate of pro-industrial hemp legislation in the United States since 1995 has dramatically increased. In 1995, there was only one state to introduce legislation, which did not pass; while in 1999 a total of nine states have passed legislation for research, study or production of the crop. States to pass legislation in 1999 are: Arkansas, California, Hawaii, Illinois, Minnesota, Montana, New Mexico, North Dakota and Virginia. The state that has taken hemp the farthest is North Dakota. On April 19, 1999 it passed the first bill in the United States that legalized hemp for commercial farming. In addition, seven states introduced legislation that did not pass in 1999. These are: Iowa, Maryland, New Hampshire, Oregon, Tennessee, Vermont and Wisconsin. Therefore, a total of sixteen states introduced legislation in 1999, representing over a third of all U.S. states. Strong pro-industrial hemp constituencies are also located in: Colorado, Idaho, Kansas, Kentucky, Missouri and Pennsylvania. See Appendix B for a listing of recent legislation.

References

Frohling, Robert E. and Staton, Eric C. 1997. "Industrial Hemp: Fertile Dream or Legal Nightmare?" NCSL LegisBrief. (Denver, CO: National Conference of State Legislatures.)

Hawaii House of Representatives. Industrial Hemp [Cannabis Sativa]- Economic Viability and Political Concerns. State of Hawaii. (Honolulu, HI) www.gametec.com/hemp/hawaii.rpt.html

Compiled by Kasea Hill, Nathan Boshard-Blackey, and Jim Simson on April 3, 2000

APPENDIX A: NCSL Legis Brief

The following lists of agricultural opportunities, economic opportunities, economic barriers and legal barriers are from the National Conference of State Legislatures brief entitled "Industrial Hemp: Fertile Dream or Legal Nightmare" and can be found at www.ncsl.org/legis/LBRIEFS/legis52.htm.

Agricultural Opportunities

The plant serves as a good rotation crop—choking out weeds and surviving without the aid of polluting pesticides, while taking no more nutrients from the soil than a corn crop.

The mature plant's strength makes it impervious to storm damage.

All parts of hemp (fiber, hurds and seeds) are economically important.

Hemp can be grown in many climates and under many conditions.

Economic Opportunities

Many industries, including those in textiles, foods, oil and building materials, have shown a growing interest in hemp. American clothing Manufacturers grossed about $5 million in 1991 and $50 million in 1995 on hemp products.

The proposed state research projects as well as the new technology and machinery needed for a developing hemp industry will provide new jobs.

Import costs for American industries using hemp, currently estimated at $120 million, would drop considerably if it could be produced domestically.

Legal Barriers

Interpretation of federal law regarding marijuana makes legislation difficult. DEA testimony in Colorado stated that they will not issue any kind of registration or permit until the federal law changes to allow industrial hemp production.

The DEA opposes industrial hemp production because it is difficult to distinguish a field of legitimate hemp, with low-narcotic concentrations, from a field of illicit cannabis with high levels. Laboratory testing is required.

The DEA fears that industrial hemp advocates have a hidden agenda to legalize marijuana. Legalization of industrial hemp could give the impression that marijuana is legal.

Economic Barriers

It may cost more money to harvest hemp. In the past, harvesting has been very labor intensive, involving no less than 11 separate operations from initial cutting to final shipping to a processing center. The plant's bulk also makes it difficult to transport.

Harvesting hemp has proved tough on today's agricultural machinery. Existing equipment must be modified to deal with the plant's rough fibers.

European hemp production has yet to prove economical. Most European governments provide substantial subsidies for growers.

Currently no one knows just how prolific hemp may be. Unlike crops such as corn, hemp has not benefited from modern research in plant genetics

APPENDIX B: Recent State Legislation

This hemp status report was prepared by Peter A. Nelson and is copyright to Agro-Tech Communications of Memphis, Tennessee. http://www.agrotechfiber.com. ©1999, Agro-Tech Communications, Memphis, Tennessee

ARKANSAS

Senator James Scott
Senate Resolution 13

On March 25, 1999 Senate Resolution 13 passed after it's third reading. This Resolution calls for the University of Arkansas to study the potential uses of Industrial Hemp and Kenaf. The Division of Agriculture will conduct studies regarding the uses and economic benefits of Industrial Hemp to determine the feasibility of growing hemp as an alternative and profitable crop in Arkansas. The studies will include an analysis of required soils and growing conditions, seed availability, harvest methods and environmental benefits. The Division of Agriculture will report its finding to the House and Senate Interim Committees on Agriculture and Economic Development no later than December 31, 2000.

CALIFORNIA

Assembly Member Strom-Martin
House Resolution 32

On September 10, 1999 House Resolution 32 passed with 41 Ayes and 30 Noes. It was resolved that the Assembly found and declared that industrial hemp is a vital sustainable, renewable resource for building materials, cloth, cordage, fiber, food, fuel, industrial chemicals, oil, paint, paper, plastics, seed, yarn, and many other useful products. It was further resolved that the Assembly found and declared that the domestic production of industrial hemp can help protect California's environment, contribute to the growth of the state economy, and be regulated in a manner that will not interfere with the enforcement of marijuana laws. It was further resolved that the Assembly found and declared that the Legislature should consider action to revise the legal status of industrial hemp to allow for its growth in California as an agricultural and industrial crop. And further resolved that the Assembly found and declared that the Legislature should consider directing the University of California, the California State University, and other state agencies to prepare studies in conjunction with private industry on the cultivation, processing, and marketing of industrial hemp. This action follows the California Democratic Party formal endorsement of a resolution supporting the development of an industrial hemp industry in California which they passed in April of 1999. In June of 1999, a resolution was proposed for adoption by California County & State Farm Bureau Federations. This resolution calls for the State of California to fund research, experimentation and development of Industrial Hemp for agricultural and industrial purposes. This work is to be conducted by the Universities of California, the California State Colleges and Universities, and other public and private companies, agencies and institutions. The resolution also calls for the (ABC) County Farm Bureau Federation to fully endorse the reintroduction of Industrial Hemp, and strongly recommend that laws be adopted by the State of California to permit the cultivation and harvesting of Industrial Hemp as a commercial crop, under the control and regulation of the California State Department of Food and Agriculture.

COLORADO

Colorado was the first state to introduce industrial hemp legislation. Bills were introduced in 1995, 1996 and 1997. Although no legislation has passed, Colorado's leadership has helped other states support US industrial hemp development.

HAWAII

Representative Cynthia Thielen

On April 8, 1999 two industrial hemp resolutions passed their final House Committee. House Resolution 122/HR109 requests that the United States Department of Agriculture, Natural Resource Conservation Service recommends the use of industrial hemp erosion control mats wherever feasible. This resolution now goes to the Senate for consideration. Resolution 123/HR110 requests that the Department of Business, Economic Development and Tourism (DBEB&T) examine the feasibility of growing industrial hemp in Hawaii for Biomass Energy Production. The DBED&T are supportive of the resolution and will complete the study regardless of further Senate action. Representative Cynthia Thielen, HB32

The Hawaii Strategic Industrial Hemp Development Act of 1999 requires the University of Hawaii at Hilo to study the feasibility and desirability of industrial hemp production in Hawaii. This bill defines "industrial hemp" and authorizes the State to allow privately funded industrial hemp research to be conducted in Hawaii. This action is pending a controlled substance registration from the State Department and a federal substance registration from the U.S. Department of Justice, Drug Enforcement Administration. On May 4, 1999, House Bill 32 passed in both the Senate and the House of Representatives. The Senate vote was 13 to 11 and all house members, except 3, voted in favor of the legislation. Governor Cayetano was supportive of the legislation and signed the bill into law on June 7, 1999. This bill utilizes a strategy introduced in Tennessee in 1998 with the Tennessee Strategic Industrial Hemp Seed Development Act of 1998 which was introduced by Representative Kathryn Bowers. This legislation is designed to utilize private funds from interested corporations in developing research programs at the state-university. The bill specifies that all agronomic data derived from research under this bill be deemed to be proprietary in nature and not subject to disclosure pursuant to the uniform information practices act. The private industry participant in the trials is expected to be Alterna Applied Research Laboratories, a California based salon products manufacturer. Unconventional in its approach, Alterna consistently sets new standards in the beauty industry in the fields of advanced formulation and product performance. The first professional hair

Company to harness the power of nutrient-rich hemp seed oil in January of 1997, Alterna continually redefines itself as an industry innovator. The field trials scheduled in Hawaii will be conducted at the College of Agriculture, Forestry and Natural Resource Management that is part of the University of Hawaii at Hilo. It is located on the Big Island of Hawaii, the largest island in the Hawaiian Archipelago. The College of Agriculture, Forestry and Natural Resource Management opened its doors in the Fall of 1975 with the main objective of preparing students with a broad and full understanding of basic factors involved in production, management, processing, distribution, marketing, sales, and services in the field of agricultural sciences including, agribusiness, animal sciences, aquaculture, agro-ecology & environmental science, crop protection, forestry, pre-veterinary medicine, sustainable agriculture, tissue culture and tropic horticulture. The College of Agriculture has approximately 130 students and 11 full-time faculty.

IDAHO

Idaho Farm Bureau Supports Industrial Hemp

On December 30, 1998 in Idaho Falls, Idaho the Idaho Farm Bureau supported the development of a US industrial hemp industry. The state group voted to adopt policy #120 which states "We encourage the legalization of cultivation and production of industrial grade hemp."

ILLINOIS

Senate Resolution 49 & House Resolution 168

Illinois Senate Resolution 49 and House Resolution 168 were passed into law at the end of March 1999. These resolutions create the Industrial Hemp Investigative and Advisory Task Force consisting of the Director of Agriculture or a designee and 12 committee members. Six members for this task force are chosen by the President of the Senate and six members are chosen by the Minority Leader of the Senate. The members of the Industrial Hemp Investigative and Advisory Task Force should represent expertise in the fields of plant science, food processing science, law enforcement, herbology, manufacturing and the Illinois Specialty Growers Association. The members of this task force will serve without compensation.

IOWA

House File 320
Representative Cecelia Burnett

Iowa's House File 320 by Representative Cecelia Burnett allows for research into industrial hemp production at Iowa State University. The bill states, "The general assembly finds that there is a trend among states to consider the economic importance of industrial hemp which is a major crop in other nations... The purpose of this Act is to promote the economy of this state by providing for research necessary to develop industrial hemp as a viable crop. Although the bill is still alive in committee it will not pass in 1999. The bill will be carried over to the 2000 legislative sessions. You can email Iowa State Representative Cecelia Burnett at cecelia_burnett@legis.state.ia.us. The 1999 legislation was rolled into the 2000 session without passing.

KANSAS

Kansas legislators have introduced legislation in 1997 and in 1998. There is a strong constituency of industrial hemp advocates in the state.

KENTUCKY

The Fayette County Farm Bureau of Lexington, Kentucky passed a resolution supporting industrial hemp in October of 1999. Actor, Woody Harrelson currently has a case before the Kentucky Supreme Court defining whether laws against marijuana in Kentucky are overly broad by including industrial hemp. On July 7, 1998 the Kentucky Hemp Growers Cooperative

Association released a landmark study in conjunction with the University of Kentucky. This study analyzed the economic potential for industrial hemp in Kentucky. The report, by the school's Center for Business and Economic Research said that cultivating and processing industrial hemp in Kentucky would bring the state up to 771 new jobs and $17.6 million in worker earnings in the current market. The Kentucky Hemp Growers Cooperative provided a public forum for legislators and all Kentucky leaders to become more informed on industrial hemp at their annual meeting in Lexington, Kentucky on June 26, 1999. Special guests included Anita Roddick, founder of The Body Shop; Jean Laprise, founder of one of Canada's largest industrial hemp growing and processing companies, Kenex Ltd., Pam Miller the Mayor of Lexington, Dr. Carl Webster of Kentucky State University and Dennis Crone a textile and agricultural fiber specialist.

MARYLAND
House Bill 374 Agriculture - Commercial Use of Industrial Hemp Act
Delegate Clarence Davis

Maryland's 1999 Commercial Use of Industrial Hemp Act authorizes the growth, maintenance, manufacture, and the regeneration of seed for the growth of industrial hemp. The bill requires the Secretary of Agriculture to develop criteria for issuing a license to engage in the commercial use of industrial hemp. This bill, introduced by Maryland Delegate Clarence Davis, received an unfavorable report from the Environmental Matters committee on March 15, 1999. The bill has not had any further action as of late April 1999.

MINNESOTA
House File 64 moved to House File 1238
Representative Kahn moved to Representative Steve Dayler

On September 30, 1999 in St. Paul, MN, Governor Ventura announced that an informational seminar would be held on November 19, 1999, to teach individual farmers how to apply for a permit to grow industrial hemp. The informational seminar is in response to legislation passed by the 1999 Minnesota Legislature requiring the state to apply by September 30 for a federal permit to grow industrial hemp. The Drug Enforcement Agency (DEA) recently informed Minnesota officials that a state cannot apply for a general blanket permit. Rather, individual farmers must apply directly to the DEA and the Minnesota Board of Pharmacy for a permit to grow experimental plots of industrial hemp. Minnesota Agriculture Commissioners Gene Hugoson and Trade and Economic Development Commissioner Jerry Carlson will co-host the November 19 seminar, giving farmers tips on how to apply for the federal and state permits they will need to grow experimental plots. They will also discuss any special conditions DEA has set for lawful cultivation of the crop. On June 4, 1999 the Minnesota Legislature passed a bill paving the way for growing experimental and demonstration plots of industrial hemp in the state, according to State Representative Phyllis Kahn (DFL-Minneapolis). Rep.Kahn was chief author of the language that was incorporated in the House Omnibus State Government Finance Bill. Governor Jesse Ventura signed this law into order on May 25, 1999. Under the new law, "by Sept. 30, 1999, the governor, in consultation with the commissioners of the Department of Agriculture and the Department of Trade and Economic Development will submit an application for federal permits, as may be needed to authorize the growing of experimental and demonstration plots of industrial hemp, by Sept. 30, 1999. The governor shall also direct the commissioner of the Department of Agriculture, in consultation with the commissioner of the Department of Public Safety and other appropriate commissioners, to establish standards and forms for persons wishing to register for growing experimental and demonstration plots of industrial hemp."

MISSOURI

The Missouri legislature considered an industrial hemp bill in 1996, 1997 and briefly in 1998. In what was virtually a one family effort, Boyd and Stacie Vancil arranged for Representative and Senate sponsorship, helped draft legislation and received backing from the Missouri Farm Bureau. The Missouri campaign matured over 1996 and 1997. Missouri quickly became the national focus for industrial hemp policy. Unfortunately, Missouri legislators were targeted by industrial hemp opponents and the pressure on the state became too much. Missouri continues to be a projected area for large-scale industrial hemp operations as federal law shifts and more states become involved. Business developments with industrial hemp in the region continue to expand.

MONTANA

House Resolution 2

Primary Sponsor: Joan Hurdle

Montana House Resolution 2 of the House of Representatives of the State of Montana requests that the federal government repeal restrictions on the production of industrial hemp as an agricultural and industrial product. The bill states "Whereas, it is a current major economic goal to diversify the agriculture of Montana; and Whereas, in over 30 countries,...,existing international treaties provide for the agricultural production and sale of industrial hemp as a valuable agricultural product; and Whereas, current federal policy is inconsistent with international agricultural policy and places an unnecessary financial restriction on the Montana agricultural community. Now therefore, be it resolved that the House of Representatives of the State of Montana urge the federal government to repeal restrictions on the production of industrial hemp as an agricultural and industrial product." This resolution is now law in the State of Montana. The House Ag Committee passed this resolution 19-0 and the House Floor passed it 95-4.

NEW HAMPSHIRE

House Bill 239
Sponsored by: Derek Owens, David Babson, Peter Leishman, Irene Messier and Amy Robb-Theroux

This bill permits the production of industrial hemp in New Hampshire. A person or business entity wishing to grow and produce industrial hemp must be licensed by the commissioner of agriculture, markets, and food. The commissioner of agriculture, markets, and food will be the sole source and supplier of seed for use in industrial hemp production.

The commissioner of agriculture, markets, and food shall charge a fee for each license granted to industrial hemp growers. The revenue from these fees is to be used to defray the costs of licensing and regulating industrial hemp growers and to fund a research program on industrial hemp production to be conducted by the University of New Hampshire. The bill grants the commissioner of agriculture, markets, and food rulemaking authority with respect to licensing and inspection of industrial hemp growers.

The 1999 legislation is still in the works! It passed a vote by the Environment and Agriculture committee in October (although it did not get an "ought to pass" recommendation by the committee) and the bill will come to the full House floor in January.

NEW MEXICO

House Bill 104
Representative Pauline K. Gubbels

New Mexico House Bill 104 makes an appropriation of Fifty thousand dollars ($50,000) for the study of industrial hemp as a commercial crop in the state. These funds are to be provided from the general fund to the board of regents of New Mexico state university for expenditure in fiscal years 2000 and 2001 for the purpose of the New Mexico department of agriculture conducting a study of the feasibility of growing industrial hemp as a commercial crop. The New Mexico department of agriculture shall report its findings to the appropriate committee during the second session of the forty-fourth legislature and first session of the forty-fifth legislature. Any unexpended or unencumbered balance remaining at the end of fiscal year 2001 shall revert to the general fund. This bill is law in New Mexico.

NORTH DAKOTA

House Bill 1428
Rep. Monson, D.Johnson, Nowatzki, Sen. Heitkamp

On Saturday, April 17, 1999 North Dakota's Governor Schafer signed HB 1428 legalizing industrial hemp by decreeing, "any person in this state may plant, grow, harvest, possess, process, sell, and buy industrial hemp."

On April 12, 1999 North Dakota's Senate passed industrial hemp bill HB1428 by a landslide vote of 44-3. The week before, the House passed the bill by 86-7. The Commissioner of Agriculture will now be developing regulations needed to implement the law which allows North Dakota farmers to legally grow industrial hemp.

The full text of the law is located at:

http://ranch.state.nd.us/LR/text/BILL_INDEX/BI1428.html

For further information contact: Gov. Ed Schafer (701) 328-2200

OREGON

House Bill 2933
Representative Prozanski

Oregon's House Bill 2933 designates definitions for industrial hemp and permits growing the crop in Oregon. On April 22, 1999 a public hearing was held on this bill. Prozanski said that seven of the nine members of the House Agriculture and Forestry Committee - including Chairman Larry Wells, R-Jefferson - had told him they were willing to send the bill out for a floor vote. But House Speaker Lynn Snodgrass told Wells not to take up the bill again. Wells, reached Thursday evening, agreed that Prozanski probably had the votes to send the bill to the floor. But, he said, he had previously assured Snodgrass he would hold just the one informational hearing on the bill, and wouldn't bring it up for a committee vote unless she approved. Snodgrass, R-Boring, could not be reached for comment Thursday evening. But Prozanski released copies of a handwritten note, written on the speaker's official letterhead, that he said Snodgrass sent to him Wednesday. The note reads, in part: "I fall back on my original feelings, am not persuaded to have the bill move forward at this time. I spoke with other members of the committee prior to making this decision." It concludes: "Keep educating the public. Perhaps future sessions are possible." Despite this motion, Prozanski's bill may not be completely kaput. Measures that pass one chamber can still be amended in the other, and proposals long since given up for dead have been known to reappear in the waning days of the session.

PENNSYLVANIA

On April 16, 1999 the forming meeting for the Pennsylvania Hemp Growers and Processor Cooperative was held in New Holland. Recently the Lancaster county's farm bureau passed a resolution to investigate the commercial potential for industrial hemp. Lancaster County Farm Bureau President, Jane Palmer said "With sinking prices for corn, soybeans and tobacco, the time is ripe for farmers to consider planting alternative crops."

RHODE ISLAND

In the 1999, TITLE 47, Weights and measures for the State of Rhode Island the state promotes legal status to industrial hemp by treating the crop as a commodity. In Section 47-4-2 of Standard Measures the following is stated:§ 47-4-2 Weights of bushels, barrels, and tons of specific commodities. – The legal weights of certain commodities in the state of Rhode Island shall be as follows: (21) A bushel of hemp shall weigh forty-four pounds (44 lbs.).

TENNESSEE

House Bill 864
Tennessee Strategic Industrial Hemp Seed Development Act
Representative Kathryn I. Bowers

The 1999 Strategic Industrial Hemp Seed Development Act would authorize agribusiness located in Tennessee to develop industrial hemp seed varieties suitable for propagation in the United States. The bill would also, subject to the approval of the commissioner of agriculture, all import of industrial hemp seed agribusiness facility. Industrial hemp is Cannabis sativa L. with a THC concentration of 1 percent or less on a dry weight basis that meets European and Canadian standards. This bill would authorize the commissioner to promulgate rules controlling the development and importation of seed. This bill is designed to work with the federal government as Federal rules are drafted over the next several years. This bill sets the groundwork for the development of agribusiness incentives for genetic research conducted in Tennessee with industrial hemp. Varieties to be developed under this act include varieties suitable to all United States regions for fiber and oil production and for international export. Tennessee House Bill 864 will be rolled into the 2000 legislative sessions as Tennessee legislator look for federal government guidance in laying out appropriate industry guidelines.

For more information concerning the 1999 Tennessee Strategic Industrial Hemp Seed Development Act, contact Peter Nelson of Agro-Tech Communications at fiber@netten.net or http://www.agrotechfiber.com.

VERMONT

Senate Bill 11 - Agriculture - Industrial Hemp
Senator Ready

This bill proposes to permit the development in Vermont of an industrial hemp industry and assure that production of industrial hemp is in compliance with state and federal laws and United States' obligations under international treaties, conventions, and protocols. Although this bill did not pass, Vermont has passed legislation in previous years that call for university study of economic and market potential of the crop.

VIRGINIA

House Joint Resolution 94
Industrial Hemp
Patrons-- Van Yahres, Bloxom,
Murphy and Wardrup; Senators: Whipple and - Woods , Mitchell Van Yahres

Summary as passed House:

Industrial hemp. Memorializes the Secretary of Agriculture, the Director of the Drug Enforcement Administration, and the Director of the Office of National Drug Control Policy to permit the controlled, experimental cultivation of industrial hemp in Virginia. Industrial hemp is seen increasingly as a potentially valuable alternative crop for farmers in Virginia, but current federal regulations make even the experimental cultivation of industrial hemp effectively impossible. The Commonwealth is also authorized to become a member of the North American Industrial Hemp Council.

WISCONSIN

Assembly Joint Resolution 49

Wisconsin AJR 49 requests that the Congress of the United States acknowledge the difference between the marijuana plant and the agricultural crop known as industrial hemp. On May 6, 1999 a public hearing was held. Wisconsin is the home of the North American Industrial Hemp Council, as well as a strong coalition of industrial hemp advocates. The Wisconsin Initiative for Industrial Hemp is endorsed by the Wisconsin National Farmers Organization, Wisconsin Agribusiness Council, Wisconsin Federation of Cooperatives, Wisconsin Farmers Union and the Wisconsin Fertilizer and Chemical Association.

APPENDIX C:

Estimates of Net Returns Per Acre for Kentucky Crops

PROCESSING TOMATOES	$ 775.0
HIGH FIBER HEMP*	$ 500.0
LOW FIBER HEMP**	$ 200.0
WHEAT AND SOYBEANS	$ 175.0
SOYBEANS	$ 100.0
HAY/SILAGE	$ 100.0
CORN	$ 75.0

* HIGH FIBER HEMP IS GROWN MORE FOR ITS FIBER.

**LOW FIBER HEMP IS GROWN MORE FOR ITS SEEDS AND HURDS THAN ITS FIBER.

REPORT TO THE (KENTUCKY) GOVERNOR'S HEMP AND RELATED FIBER CORP.

ARTICLE SOURCE: VERMONT LEGISLATIVE RESEARCH SHOP VIABILITY OF INDUSTRIAL HEMP

WWW.VOTEHEMP.COM/PDF/VLRS_HEMP.PDF

Hemp Oil Fuels & How to Make Them

http://website.lineone.net/~supersnowy1/Medlab_AplicationsoilFuel.htm

By: A. Das

Introduction of Hemp Biodiesel maybe the liquid fuel of the future. Hemp is a high yield C-4 photosynthesis plant. Hemp can Boost a higher oilseed yield than any of today's oilseed crops (soy, canola or safflower).

Thirty years ago soy beans were a joke to American farmers. Who would have guessed that in thirty years soy beans would become the largest oil and protein crop in American farming. Right now Hemp farming is a joke to American farmers. Who knows what the next thirty years will do to <u>American Hemp farming.</u>

Hemp fuels are yet another benefit of Domestic Industrial Hemp Farming. As we enter 1997 more than ten states will be considering Industrial Hemp Farm Bills. In the mean times Hempseed must be grown out side the country. The major part of the cost of the inexpensive hempseed is transportation from across the globe. The seed to produce a gallon of hempseed oil can cost up to $100. All foreign production and shipping plans are doomed to high costs. I look forward to the days when a farmer can produce his own hempseed oil fuel as low as a dollar a gallon.

The following formula for making Hemp Diesel Fuel will work nicely to make small quantities of fuel to run the sound stage at your Hemp Rally this summer. A 4 kilowatt diesel generator uses around one litre an hour .Imagine walking to the microphone and saying, "The sound of my voice is coming to you with the power of Hemp Fuel !". Seeing is believing. I'll drive you around the state capital, Senator, In my Hemp Fueled Vehicle!

Bio Diesels not a new fuel. The DOE and USDA have provided funding for research for years. The Biomass Conference of The Americas in Burlington Vermont had over a dozen papers presented on all aspects of Hemp as an oilseed cultivars. Let's get on with it!

Hemp, four times more efficient than corn as biofuel, is a clean, smart and economical energy choice.

How to Make Bio Diesel
By: A. Das

CAUTION!

TITRATION OF FREE FATTY ACIDS

Measure Free Fatty Acid content of your oil: Mix 1 ml oil with 10 ml Isopropyl alcohol = 2 drops phenolthalian solution (available in a hobby shop chemistry set suppliers). Drop wise add 0.1% lye solution (1 gm lye in one litre water) until the solution stays pink for 10 seconds. (20 drops = 1 ml) Record the millilitres of 0.1% lye solution used.

METHANOL

You will need 200 ml of methanol per litre of Hemp Seed oil. Methanol may be purchased as Drigas available at most automotive stores, read the label for methanol. Also Methanol is available from racing stores. Avoid hardware store methanol (wood alcohol) as it may contain excessive water content.

SODIUM METHOXIDE

For each liter of hemp seed oil you need one gram of granular solid lye for each ml of 0.1% lye solution used in titration of free fatty acids plus 3.5 grams. Completely dissolve the proper amount of Lye in the methanol (Red Devil Lye can be purchased from the Grocery Store). This combined mixture makes sodium methoxide.

MIXER

The type of mixer depends on the size of the batch. A blender works fine for a small batch. An electric drill and paint mixer on an extended shaft works well in a 5 gallon bucket. An electric light dimmer switch provides a good speed control.

TRANSESTERFICATION

Once the lye catalyst is dissolved completely so that there is no sediment, then the oil may be added to the methanol lye mixture while mixing continuously. At first the mixture becomes thicker, *then thinner as the reaction proceeds. Collect samples every 5 minutes with an eye dropper into a test tube or clear container. The Mixture will separate into a light top layer of bio diesel and a darker bottom layer of glycerin, soap and catalyst. Continued mixing 30 - 60 minutes until the yield remains constant. Then stop mixing. Go have lunch.*

When you come back it will have settled into two distinct layers. You have just made what could be the fuel of the future for a self reliant society. Let the mixture settle for at least 8 hours. Pour off and save the bio diesel top layer into another container. A clear funnel bottomed container is helpful.

RINSING

The raw biodiesel that you have just produced may have some catalyst, alcohol, and glycerin remaining which could cause engine problems, so for long term engine reliability this raw fuel should be rinsed with water. Gently at first then more vigorously rinse with water until the rinse water is clear and the pH of the rinse water is the same pH as the supply water. Settle, Decant.

DRYING

Water in the bio Diesel makes cloudy so it must be carefully heated. At 100 C most of the water coalesces and falls to the bottom. This water must be completely removed from the bottom of the container before heating to higher temperature.

CAUTION!

WEAR PROTECTIVE CLOTHING AND EYEWEAR.

FAILURE TO REMOVE THIS WATER BEFORE FURTHER HEATING CAN CAUSE VIOLENT ERUPTION OF HOT LIQUID!

Once all water has been removed then heat the bio diesel to 300 f (150 c) to complete dryness. Cool, filter, and store bio diesel in a well marked dry closed container. 100% **HEMP DIESEL FUEL (HEMP OIL METHYL ESTER - HOME FUEL)**

This fuel may be mixed in any ratio with petroleum diesel. Dynamometer tests indicate full power output with up to 75% reduction in soot and particulates. No engine modification is needed to burn bio diesel fuel.

Other Oil Feed Stocks

By: A. Das

Hemp Seed Oil at present is too expensive to drive across the country. That is not the object of this article. Our propose is to demonstrate proof of feasibility of this fuel concept. The time is now to give hemp a chance. The small quantities of Hemp Diesel Fuel can play a powerful role in educating ourselves and the policy makers about the hope in hemp.

For other readers the question will be raised. What else can I use can I use as a feedstock that is cheaper between now and domestic hemp seed crops? Soy, Sunflower, Canola, and Safflower oils are being used in field testing programs right now. The ground support vehicles at the Kansas City Airport are operating on Soya diesel. Lincoln Nebraska City busses are operating on Bio Diesel.

Go ahead practice your fuel making technique on any vegetable oil available. The most important change for us may start within ourselves. Let us get on with the curriculum.

Fat Of The Land

A four woman video crew recently traveled across the country from new York city to San Francisco in a Diesel Chevy Van fueled by French Fryer Bio Diesel. They would drive past the gas pumps and the diesel pumps. They would drive around the back of the burger joint and ask for drippings from the fryer in day-glow pink waitress outfits. The video is both entertaining and informative. Video is both available from Original Sources (303) 237 - 3579.

SEVEN WAYS AROUND THE GAS PUMP USING HEMP FUEL

BY: A. DAS

A BOOK IS FORTH COMING ON SEVEN WAYS TO RUN YOUR CAR USING HEMP STALKS AND SEEDS.

Text of the Yearbook of
The United States Department Of Agriculture—1913, pages 283-346

HEMP

BY: LYSTER H. DEWEY
BOTANIST IN CHARGE OF FIBER PLANT INVESTIGATIONS
BUREAU OF PLANT INDUSTRY

INTRODUCTION

The two fiber-producing plants most promising for cultivation in the central United States and most certain to yield satisfactory profits are hemp and flax. The oldest cultivated fiber plant, one for which the conditions in the United States are as favorable as anywhere in the world, one which properly handled improves the land, and which yields one of the strongest and most durable fibers of commerce, is hemp. Hemp fiber, formerly the most important material in homespun fabrics, is now most familiar to the purchasing public in this country in the strong gray tying twines one-sixteenth to one-fourth inch in diameter, known by the trade name "commercial twines."

NAME

The name "hemp" belongs primarily to the plant Cannabis sativa. (pl. XL, fig. 1.) It has long been used to designate also the long fiber obtained from the hemp plant. (Pl. XL, fig. 4.) Hemp fiber, being one of the earliest and best-known textile fibers and until recent times the most widely used of its class, has been regarded as the typical representative of long fibers. Unfortunately, its name also came to be regarded as a kind of common name for all long fibers, until one now finds in the market quotations "Manila hemp" for abaca, "sisal hemp" for sisal and henequen, "Mauritius hemp" for Furcraea fiber, "New Zealand hemp" for phormium, "Sunn hemp" for Crotalaria fiber, and "India hemp" for jute. All of these fibers in appearance and in economic properties are unlike true hemp, while the name is never applied to flax, which is more nearly like hemp than any other commercial fiber.

The true hemp is known in different languages by the following names: Cannabis, Latin; chanvre, French; canamo, Spanish; canhamo, Portuguese; canapa, Italian; canep, Albanian; konopli, Russian; konopj and penek, Polish; kemp, Belgian; hanf, German; hennep, Dutch; hamp, Swedish; hampa, Danish; kenevir, Bulgarian; ta-ma, si-ma, and tse-ma, Chinese; asa, Japanese; nasha, Turkish; kanabira, Syrian; kannab, Arabic.

IMPORTANCE OF HEMP

Hemp was formerly the most important long fiber, and it is now used more extensively than any other soft fiber except jute. From 10,000 to 15,000 tons are used in the United States every year. The approximate amount consumed in American spinning mills is indicated by the following table, showing the average annual importations (Computed from reports of the Bureau of Navigation and Commerce, U.S. Treasury Department Bureau of Statistics Department of Commerce) and estimates of average domestic production of hemp fiber for 35 years:

Average annual imports and estimates of average annual production of hemp fiber in 5-year periods from 1876 to 1910, inclusive, and from 1911 to 1913, inclusive. (Missing chart from p. 284)

There are no statistics available, such as may be found for wheat, corn, or cotton, showing with certainty the acreage and production of hemp in this country. The estimates of production in the foregoing table are based on the returns of the Commissioner of Agriculture of Kentucky for earlier years with amounts added to cover the production in other States, and on estimates of hemp dealers for more recent years. While these figures can not be regarded as accurate statistics, and they are probably below rather than above the actual production, especially in the earlier years, they indicate a condition well recognized by all connected with the industry. The consumption of hemp fiber has a slight tendency to increase, but the increase is made up through increased importations, while the domestic production shows a tendency toward reduction.

PRODUCTION IN THE UNITED STATES DECLINING

This falling off in domestic production has been due primarily to the increasing difficulty in securing sufficient labor to take care of the crop; secondarily, to the lack of development of labor-saving machinery as compared with the machinery for handling other crops and to the increasing profits in raising stock, tobacco, and corn, which have largely taken the attention of farmers in hemp-growing regions. The work of retting, breaking, and preparing the fiber for market requires a special knowledge, different from that for handling grain crops, and a skill best acquired by experience. These factors have been more important than all others in restricting the industry to the bluegrass region of Kentucky, where the plantation owners as well as the farm laborers are familiar with every step in handling the crop and producing the fiber.

An important factor, tending to restrict the use of hemp, has been the rapidly increasing use of other fibers, especially jute, in the manufacture of materials formerly made of hemp. Factory-made woven goods of cotton or wool, more easily spun by machinery, have replaced the hempen "homespun" for clothing; wire ropes, stronger, lighter, and more rigid, have taken its place in standing rigging for ships; abaca (Manila hemp), lighter and more durable in salt water; has superseded it for towing hawsers and hoisting ropes; while jute, inferior in strength and durability, and with only the element of cheapness in its favor, is usurping the legitimate place of hemp in carpet warps, so-called "hemp carpets," twines, and for many purposes where the strength and durability of hemp are desired. The introduction of machinery for harvesting hemp and also for preparing the fiber, together with the higher prices paid for hemp during the past three years, has aroused an interest in the industry, and many experiments are being tried with a view to the cultivation of the crop in new areas.

BOTANICAL STUDY OF HEMP

THE PLANT

The hemp plant, Cannabis sativa L., (Linnaeus. Species Plantarum, ed. 1,1027, 1753. Dioscorides. Medica Materia, libri sex, p. 147, 1537. Synonyms: Cannabis erratica paludesa Anders. Lobel. Stirpium Historia, 184, 1576. Cannabis indica Lamarck. Encyclopedia, 1: 695, 1788. Cannabis macrosperma Stokes. Bot. Mat. Med., IV, 539, 1812. Cannabis chinensis Delile. Ind. Sem. Hort. Monst. in Ann. Sci. Nat. Bot., 12: 365, 1849. Cannabis gigantea Delile. L. Vilmorin. Rev. Hort., 5: s. 3, 109, 1851.) is an annual, growing each year from the seed. It has a rigid, herbaceous stalk, attaining a height of 1 to 5 meters (3 to 16 feet), obtusely 4-cornered, more or less fluted or channeled, and with well-marked nodes at intervals of 10 to 50 centimeters (4 to 20 inches). When not crowded it has numerous spreading branches, and the central stalk attains a thickness of 3 to 6 centimeters (1 to 2 inches), with a rough bark near the base. If crowded, as when sown broadcast for fiber, the stalks are without branches or foliage except at the top, and the smooth fluted stems are 6 to 20 millimeters (1/4 to 3/4 inch) in diameter. The leaves, opposite, except near the top or on the shortened branches, appearing fascicled, are palmately compound and composed of 5 to 11--usually 7--leaflets. (Pl. XLI, fig. 1). The leaflets are dark green, lighter below, lanceolate, pointed at both ends, serrate, 5 to 15 centimeters (2 to 6 inches) long, and 1 to 2 centimeters (3/8 to 3/4 inch) wide. Hemp is dioecious, the staminate or pollen-bearing flowers and the pistillate or seed-producing flowers being borne on separate plants. The staminate flowers (Pl. XL, fig. 2) are borne in small axillary panicles, and consist of five greenish yellow or purplish sepals

182

opening wide at maturity and disclosing five stamens which discharge abundant yellow pollen. The pistillate flowers (Pl. XL, fig. 3) are stemless and solitary in the axils of the small leaves near the ends of the branches, often crowded so as to appear like a thick spike. The pistillate flower is inconspicuous, consisting of a thin, entire, green calyx, pointed, with a slit at one side, but remaining nearly closed over the ovary and merely permitting the two small stigmas to protrude at the apex. The ovary is one seed, developing into a smooth, compressed or nearly spherical achene (the "seed"), 2.5 to 4 millimeters (1/10 to 3/16 inch) thick and 3 to 6 millimeters (1/8 to 1/4 inch) long, from dark gray to light brown in color and mottled (Pl. XLI, fig. 2). The seeds cleaned for market nearly always include some still covered with the green, gummy calyx. The seeds vary in weight from 0.008 to 0.027 gram, the dark-colored seeds being generally much heavier than the light-colored seeds of the same sample. The light-colored seeds are often imperfectly developed. Dark-colored and distinctly mottled seeds are generally preferred. The staminate plants are often called the flowering hemp, since the pistillate flowers are rarely observed. The staminate plants die after the pollen is shed, but the pistillate plants remain alive and green two months later, or until the seeds are fully developed.

THE STALK

The hemp stalk is hollow, and in the best fiber-producing types the hollow space occupies at least one-half the diameter. The hollow space is widest, or the surrounding shell thinnest, about midway between the base and the top of the plant. The woody shell is thickened at each node, dividing the hollow space into a series of partly separated compartments. (Pl. XLI, fig. 4.) If the stalk is cut crosswise a layer of pith, or thin-walled tissue, is found next to the hollow center, and outside of this a layer of wood composed of hard, thick-walled cells. This layer, which forms the "hurds," is a very thin shell in the best fiber-producing varieties. It extends clear across the stem below the lowest node, and in large, coarse stalks grown in the open it is much thicker and the central hollow relatively smaller. Outside of the hard woody portion is the soft cambium, or growing tissue, the cells of which develop into the wood on the inside, or into the bast and bark on the outside. It is chiefly through this cambium layer that the fiber-bearing bast splits away from the wood in the processes of retting and breaking. Outside of the cambium is the inner bark, or bast, comprising short, thin-walled cells filled with chlorophyll, giving it a green color, and long thick-walled cells, making the bast fibers. These bast fibers are of two kinds, the smaller ones (secondary bast fibers) toward the inner portion making up rather short, fine fibers, many of which adhere to the wood or hurds when the hemp is broken, and the coarser ones (primary bast fibers) toward the outer part, extending nearly throughout the length of the stalk. Outside of the primary bast fiber is a continuation of the thin-walled chlorophyll-bearing cells free from fiber, and surrounding all is the thin epidermis.

**Hemp Fibers
1913 Yearbook**

THE FIBER

The hemp fiber of commerce is composed of the primary bast fibers, with some adherent bark and also some secondary bast fiber. The bast fibers consist of numerous long, overlapping, thick-walled cells with long, tapering ends. The individual cells, almost too small to be seen by the unaided eye are 0.015 to 0.05 millimeter (3/1000 to 12/1000 inch) in diameter, and 5 to 55 millimeters (3/16 to 2 1/8 inches) long. Some of the bast fibers extend through the length of the stalk, but some are branched, and some terminate at each node. They are weakest at the nodes.

RELATIONSHIPS

The hemp plant belongs to the mulberry family, Moraceae, which includes the mulberry, the Osage orange, the paper mulberry, from the bast of which the tapa of the South Sea Islands is made, and the hop, which contains a strong bast fiber. Hemp is closely related to the nettle family, which includes ramie, an important fiber-producing plant of Asia and several species of nettles having strong bast fibers. The genus Cannabis is generally regarded by botanists as monotypic, and the one species Cannabis sativa is now held to include the half dozen forms which have been described under different names (see footnote, p. 286) and which are cultivated for different purposes. The foregoing description refers especially to the forms cultivated for the production of fiber.

HISTORY

EARLY CULTIVATION IN CHINA

Hemp was probably the earliest plant cultivated for the production of a textile fiber. The "Lu Shi," a Chinese work of the Sung dynasty, about 500 A.D., contains a statement that the Emperor Shen Nung, in the twenty-eighth century B.C., first taught the people of China to cultivate "ma" (hemp) for making hempen cloth. The name ma (fig. 17) occurring in the earliest Chinese writings designated a plant of two forms, male and female, used primarily for fiber. Later the seeds of this plant were used for food. (- Bretschneider, E. Botannicum

麻

Sinicum, in Journal of the North *China Branch of the Royal Asiatic Society, n.s., v. 25, p.203, 1893, Shanhai.) The definite statement regarding the staminate and pistillate forms eliminates other fiber plants included in later times under the Chinese name ma. The Chinese have cultivated the plant for the production of fiber and for the seeds, which were used for food and later for oil, while in some places the stalks are used for fuel, but there seems to be no record that they have used the plant for the production of the narcotic drugs bhang, charas, and ganga. (sic)` The production and use of these drugs were developed farther west.*

CULTIVATION FOR NARCOTIC DRUGS

The use of hemp in medicine and for the production of the narcotic drug Indian hemp, or cannabis, is of interest in this paper only because of its bearing on the origin and development of different forms of the plant. The origin of this use is not definitely known, but the weight of evidence seems to indicate central Asia or Persia and a date many centuries later than its first cultivation for fiber. The name bhanga occurs in the Sanskrit "Atharvaveda" (about 1400 B.C.), but the first mention of it as a medicine seems to be in the work of Susruta (before the eighth century A.D.), while in the tenth century A.D. its intoxicating nature seems to have been known, and the name "indracana" (Indra's food) first appears in literature. (Watt, Sir George. Commerical Products of India, p. 251, 1908.) A further evidence that hemp, for the production of fiber as well as the drug, has been distributed from central Asia or Persia is found in the common origin of the names used. The Sanskrit names "bhanga" and "gangika," slightly modified to "bhang" and "ganja," are still applied to the drugs, and the roots of these words, "and" and "an," recur in the names of hemp in all of the Indo-European and modern Semitic languages, as bhang, ganja, hanf, hamp, hemp, chanvre, canamo, kannab, cannabis. (De Candolle, Alphonse. Origin of Cultivated Plants, p. 143, 1886)

HEMP IN INDIA

Northern India has been regarded by some writers as the home of the hemp plant, but it seems to have been unknown in any form in India before the eighth century, and it is now thought to have been introduced there first as a fiber plant. It is still cultivated to a limited extent for fiber in Kashmir and in the cool, moist valleys of the Himalayas, but in the warmer plains regions it is grown almost exclusively for the production of the drugs. (Watt, Sir George. Commercial Products of India, p. 253, 1908.) Hemp was not known to the Hebrews nor to the ancient Egyptians, but in medieval times it was introduced into North Africa, where it has been cultivated only for the drug. It is known in Morocco as "kif," and a small form, 1 to 3 feet high, cultivated there has been described as a distinct variety, Cannabis sativa kif. (De Candolle, Alphonse. Prodromus, v. 16, pt. 1, p.31,1869.)

INTRODUCTION INTO EUROPE

According to Herodotus (about 450 B.C.), the Thracians and Scythians, beyond the Caspian Sea, used hemp, and it is probable that the Scythians introduced the plant into Europe in their westward migration, about 1500 B.C., though it seems to have remained almost unknown to the Greeks and Romans until the beginning of the Christian era. The earliest definite record of hemp in Europe is the statement that "Hiero II, King of Syracuse (270 B.C.), bought hemp in Gaul for the cordage of his vessels." (De Candolle, Alphonse. Origin of Cultivated Plants, p. 148, 1886.) From the records of Tragus (1539 A.D.), hemp in the sixteenth century had become widely distributed in Europe. It was cultivated for fiber, and its seeds were cooked with barley and other grains and eaten, though it was found dangerous to eat too much or too frequently. Dioscorides called the plant Cannabis sativa, a name it has continued to bear to the present time, and he wrote of its use in "making the stoutest cords" and also of its medicinal properties. ((Dioscorides. Medica Materia, li bri sex, p. 147, 1537.) Nearly all of the early herbalists and botanical writers of Europe mention hemp, but there is no record of any further introduction of importance in the fiber industry until the last century.

INTRODUCTION OF CHINESE HEMP INTO EUROPE

In 1846 M. Hebert sent from China to the Museum at Paris some seeds of the "tsing-ma" great hemp, of China. Plants from this seed, grown at Paris by M.L. Vilmorin, attained a height of more than 15 feet, but did not produce seeds. In the same year M. Itier sent from China to M. Delile, of the Garden at Montpellier, France, seeds of a similar kind of hemp. These seeds were distributed in the southern part of France, where the plants not only grew tall, some of them measuring 21 feet, but they also produced mature seeds. M. Delile called this variety Cannabis chinensis (Delile, Raffenau. Index seminum hortl botanici Monspeliensis. Ann. Sci. Nat. Bot., v. 12, p. 365, 1849.) and the one from the seeds sent by M. Hebert he called C. gigantea. (Vilmorin, L. Chanvre de Chine. Rev. Hort. 5: s. 3, p. 109, 1851) These two forms of hemp were regarded as the same by M.L. Vilmorin, who states that they differ very much in habit from the common hemp of Europe, which was shorter and less valuable for fiber production. We are also told that this chanvre de Chine did not appear to be the same as the chanvre de Piedmont, (5-footnote missing from transcription, p. 290) the tall hemp of eastern France and northern Italy, the origin of which has sometimes been referred to this introduction, but this may have originated in a previous introduction, since Cannabis chinensis is mentioned as having been in the Botanical Garden at Vienna in 1827. In the same statement, however, C. sativa pedemontana is described as a distinct variety. (De Candolle, Alphonse. Prodromus, v. 16, pt. 1, p.31, 1869.) Particular attention is called to the introduction of this large Chinese hemp into Europe, since it was doubtless from the same source as the best hemp seed now brought from China to the United States.

186

INTRODUCTION INTO SOUTH AMERICA

Hemp from Spain was introduced into Chile about 1545. (Husbands, Jose D. U.S. Department of Agriculture, Bureau of Plant Industry, Bulletin 153, p. 42, 1909.) It has been largely grown in that country, but at present its cultivation is confined chiefly to the fertile lands in the valley of the Rio Aconcagua, between Valparaiso and Los Andes, where there are large cordage and twine mills. The fiber is all consumed in these mills.

INTRODUCTION INTO NORTH AMERICA

Hemp was introduced into New England soon after the Puritan settlements were established, and the fact that it grew "twice so high" as it did in old England was cited as evidence of the superior fertility of the soil of New England. (Morton, Thomas. New English Canaan, p. 64, 1632. In Force, Peter, Tracts and Other Papers, v. 2, 1838.) A few years later a writer in Virginia records the statement that "They begin to plant much Hempe and Flax which they find growes well and good." (Virginia, printed for Richard Wodenoth, 1649. In Force, Peter, Tracts and Other Papers, v. 2, 1838.) The cultivation of hemp in the New England colonies, while continued for some time in Massachusetts and Connecticut, did not attain as much importance as the cultivation of flax for supplying fiber for household industry. In the South hemp received more attention, especially from the Virginia Legislature, which passed many acts designed to promote the industry, but all in vain. (Moore, Brent. A Study of the Past, the Present, and the Possibilities of the Hemp Industry in Kentucky, p. 14, 1905.)

The cultivation of hemp seems to have been a flourishing industry in Lancaster County, Pa., before the Revolution. An elaborate account of the methods then employed in growing hemp, written about 1775 by James Wright, of Columbia, Pa., (New Era, Lancaster, Pa., June 25, 1905.) was recently published as an historical document. The methods described for preparing the land were equal to the best modern practice, but the hemp was pulled by hand instead of cut. Various kinds of machine brakes had been tried, but the had all "given Way to one simple Break of a particular Construction, which was first invented & made Use of in this country." The brief description indicates the common hand brake still in use in Kentucky.

USDA Workers With Hemp Bundles 1913 Yearbook

EARLY CULTIVATION IN KENTUCKY

The first crop of hemp in Kentucky was raised by Mr. Archibald McNeil, near Danville, in 1775. (Moore, Brent. A Study of the Past, the Present, and the Possibilities of the Hemp Industry in Kentucky, p. 16, 1905.) It was found that hemp grew well in the fertile soils of the bluegrass country, and the industry was developed there to a greater extent than it had been in the eastern colonies. While it was discontinued in Massachusetts, Virginia, and Pennsylvania, it has continued in Kentucky to the present time. In the early days of this industry in Kentucky, fiber was produced for the homespun cloth woven by the wives and daughters of the pioneer settlers, and an export trade by way of New Orleans was developed. In 1802 there were two extensive ropewalks in Lexington, Ky., and there was announced "a machine, moved by a horse or a current of water, capable, according to what the inventor said, to break and clean eight thousand weight of hemp per day." (Michaux, F. Andre. Travels to the west of the Alleghanies, p. 152, 1805. In Thwaites, Early Western Travels, v. 3, p. 200, 1904.) Hemp was later extensively used for making cotton-bale covering. Cotton bales were also bound with hemp rope until iron ties were introduced, about 1865. There was a demand for the better grades of hemp for sailcloth and for cordage for the Navy, and the industry was carried on more extensively from 1840 to 1860 than it has been since.

EXTENSION OF THE INDUSTRY TO OTHER STATES

Hemp was first grown in Missouri about 1835, and in 1840 1,600 tons were produced in that State. Four years later the output had increased to 12,500 tons, and it was thought that Missouri would excel Kentucky in the production of this fiber. With the unsatisfactory methods of cleaning the fiber on hand brakes and the difficulties of transporting the fiber to the eastern markets, hemp proved less profitable than other crops, and the industry was finally abandoned about 1890. Hemp was first grown at Champaign, Ill., about 1875. A cordage mill was established there for making twines from the fiber, which was prepared in the form of long tow by a large machine brake. The cordage mill burned and the industry was discontinued in 1902 because there was no satisfactory market for the kind of tow produced.

In Nebraska, hemp was first grown at Fremont in 1887 by men from Champaign, Ill. A binder-twine plant was built, but owing to the low price of sisal, more suitable for binder twine, most of the hemp was sold to eastern mills to be used in commercial twines. After experimenting with machine brakes the company brought hand brakes from Kentucky and colored laborers to operate them. The laborers did not stay, and the work was discontinued in 1900. Some of the men who had been connected with the company at Fremont began growing hemp at Havelock, near Lincoln, in 1895. A machine for making long tow, improved somewhat from the one at Champaign, was built. Further improvements were made in the machine and also in the methods of handling the crop, but the industry was discontinued in 1910, owing to the lack of a satisfactory market for the kind of tow produced. Hemp was first grown on a commercial scale in California at Gridley, in Butte County, by Mr. John Heaney, who had grown it at Champaign and who devised the machine used there for making long tow. Mr. Heaney built a machine with some improvements at Gridley, and after three disastrous inundations from the Feather River moved to Courtland, in the lower Sacramento Valley, where the reclaimed lands are protected by dikes. The work is now being continued at Rio Vista, in Solano County, under more favorable conditions and with a machine still further improved. The hemp fiber produced in California is very strong and is generally lighter in color than that produced in Kentucky. In 1912 hemp was first cultivated on a commercial scale under irrigation at Lerdo, near Bakersfield, Cal., and a larger acreage was grown there in 1913. The seed for both crops was obtained in Kentucky.

INTRODUCTION OF CHINESE HEMP INTO AMERICA

In 1857 the first Chinese hemp seed was imported. It met with such favor that some of this seed is said to have brought $10 per quart. (Moore, Brent. The Hemp Industry in Kentucky, pp. 60-61, 1905.) Since that time the common hemp of European origin has given place in this country to the larger and better types from China.

GEOGRAPHICAL DISTRIBUTION

The original home of the hemp plant was in Asia, and the evidence points to central Asia, or the region between the Himalayas and Siberia. Historical evidence must be accepted rather than the collection of wild specimens, for hemp readily becomes naturalized, and it is now found growing without cultivation in all parts of the world where it has been introduced. Hemp is abundant as a wild plant in many localities in western Missouri, Iowa, and in southern Minnesota, and it is often found as a roadside weed throughout the Middle West. De Candolle (De Candolle, Alphonse. Origin of Cultivated Plants, p. 148, 1886.) writes of its origin as follows:

The species has been found wild, beyond a doubt, south of the Caspian Sea (De Bunge); in Siberia, near the Irtyach; and in the Desert of Kirghiz, beyond Lake Baikal, in Dahuria (Government of Irkutsh). It is found throughout central and southern Russia and south of the Caucasus, but its wild nature here is less certain. I doubt whether it is indigenous in Persia, for the Greeks and Hebrews would have known of it earlier.

Hemp is now cultivated for the production of fiber in China, Manchuria, Japan, northern India, Turkey, Russia, Austria-Hungary, Italy, France, Belgium, Germany, Sweden, Chile, and in the United States. It is grown for the production of the drugs bhang, ganja, kif, marihuana, hasheesh, etc., in the warm, arid, or semiarid climates of India, Persia, Turkey, Algeria, central and southern Africa, and in Mexico, and for the production of seed for oil in China and Manchuria. In the United States hemp is now cultivated in the bluegrass region of Kentucky within a radius of 50 miles of Lexington; in the region of Waupun, Wis.; in northern Indiana; near Lima, Ohio; and at Lerdo and Rio Vista, Cal. There are numerous small experimental plats in other places. The principal countries producing hemp fiber for export are Russia, Italy, Hungary, and Roumania. China and Japan produce hemp fiber of excellent quality, but it is nearly all used for home consumption. Hemp is not cultivated for fiber in the Tropics or in any of the warm countries. The historical distribution of hemp, as nearly as may be traced from the records, and the areas where hemp is now cultivated are indicated in the accompanying map, figure 6.

VARIETIES

Hemp, cultivated for three different products--fiber from the bast, oil from the seeds, and resinous drugs from the flowers and leaves--has developed into three rather distinct types or groups of forms. The extreme, or more typical, forms of each group have been described as different species, but the presence of intergrading forms and the fact that the types do not remain distinct when cultivated under new conditions make it impossible to regard them as valid species. There are few recognized varieties in either group. Less than 20 varieties of fiber-producing hemp are known, although hemp has been cultivated for more than 40 centuries, or much longer than either cotton or corn, both of which now have hundreds of named varieties.

CHINA

The original home of the hemp plant was in China, and more varieties are found there than elsewhere. It is cultivated for fiber in nearly all parts of the Chinese Republic, except in the extreme south, and over a wide range of differences in soil and climate with little interchange of seed, thus favoring the development and perpetuation of varietal differences. The variety called "ta-ma" (great hemp) is cultivated chiefly in the provinces of Chekiang, Kiangsu, and Fukien, south of the Yangtze. In the rich lowland soils, often in rotation with rice, but not irrigated, and with a warmer and longer growing season than in Kentucky, this hemp attains a height of 10 to 15 feet. The seed is dark colored, usually well mottled, small, weighing about 1.2 grams per hundred. The internodes of the main stem are 6 to 10 inches long; the branches long and slender, usually drooping at the ends; the leaves large; and the pistillate flowers in small clusters. Seed brought from China to Kentucky in recent years is mostly of this variety. When first introduced it is too long in maturing to permit all of the seeds to ripen. The most important fiber plant of western China is the variety of hemp called "hoa-ma." It is grown in the province of Szechwan and as a winter crop on the plains of Chengtu in that province. It is shorter and more compact in its habit of growth and earlier in maturing than the ta-ma of the lowlands.

190

A variety called "shan-ma-tse" is cultivated in the mountain valleys in the provinces of Shansi and Chihli, in northern China. Its fiber is regarded as the best in North China, and in some respects as superior to that of ta-ma, though the yield is usually smaller. The plants attain a height of 6 to 9 feet, with a very thin woody shell, short ascending branches, rather small leaves, and larger seeds in larger clusters than those of ta-ma. Imported seed of this variety, grown in a trial plat in Kentucky, produced plants smaller in size and maturing earlier than Kentucky hemp. In the mountains both north and south of Ichang in central China a variety called "t'ang-ma" (cold hemp) is cultivated primarily for the production of seeds, from which oil is expressed. It is a very robust form, with stalks 6 to 12 feet high and 2 to 4 inches in diameter. These stalks are used for fuel, and occasionally a little fiber is stripped off for domestic use.

[Fig. 18 Map in original]

In Manchuria two distinct kinds of hemp are cultivated. One, called "hsien-ma," very similar to the shan-ma-tse of northern China, is grown for fiber. It attains a height of 8 to 9 feet, and requires nearly 150 days from seeding to full maturity. The other, called "shem-ma," is grown for oilseed production. It attains a height of 3 to 5 feet and is ripe with fully matured seeds in less than 100 days. The branches usually remain undeveloped, so that the clusters of seeds are borne in compact heads at the tops of the simple stalks. (Pl. XLII, fig.1.) It is said that in Manchuria these two forms remain distinct without crossing or producing any intergrading forms. The Chinese name "ma" (fig. 17), originally applied only to the true hemp (Cannabis sativa), is now used as a general term to designate nearly all textile plants in China. (Bretschneider, E. Botanicum Sinicum, p. 203, 1893.) This general use leads to nearly as much confusion among English-speaking people in China as does the unfortunate use of the name hemp as a synonym for fiber in this country. The staminate hemp plant is called "si-ma," and the pistillate plant "tsu-ma." Flax, cultivated to a limited extent in northern China, is called "siao-ma" (small hemp), but this name is also applied to small plants of true hemp. Ramie, cultivated in central and southern China, is "ch'u-ma" or "tsu-ma." China jute, cultivated in central and northern China and in Manchuria and Chosen (Korea), is called "tsing-ma," or "ching-ma," and its fiber, exported from Tientsin, is called "pei-ma." India jute, cultivated in southern China and Taiwan, is called "oi-ma." The name "chih-ma" is also applied in China to sesame, which is not a fiber plant.

JAPAN

Hemp, called "asa" in the Japanese language, is cultivated chiefly in the provinces or districts of Hiroshima, Tochigi, Shimane, Iwate, and Aidzu, and to a less extent in Hokushu (Hokkaido) in the north and Kiushu in the south. It is cultivated chiefly in the mountain valleys, or in the north on the interior plains, where it is too cool for cotton and rice and where it is drier than on the coastal plain. That grown in Hiroshima, in the south, is tall, with a rather coarse fiber; that in Tochigi, the principal hemp-producing province, is shorter, 5 to 7 feet *high, with the best and finest fiber, and in Hokushu it is still shorter. Seeds from Hiroshima, Shimane, Aidzu, Tochigi, and Iwate were tried by the United States Department of Agriculture in 1901 and 1902. The plants showed no marked varietal differences. They were all smaller than the best Kentucky hemp. The seeds varied from light grayish brown, 5 millimeters (1/5 inch) long, to dark gray, 4 millimeters (1/6 inch) long. The largest plants in every trial plat were from Hiroshima seeds, and these seeds were larger and lighter colored than those of any other variety except Shimane, the seeds of which were slightly larger and the plants slightly smaller.*

RUSSIA

Hemp is cultivated throughout the greater part of Russia, and it is one of the principal crops in the provinces of Orel, Kursk, Samara, Smolensk, Tula, Voronezh, and Poland. Two distinct types, similar to the tall fiber hemp and the short oil-seed hemp of Manchuria, are cultivated, and there are doubtless many local varieties in isolated districts where there is little interchange of seed. The crop is rather crudely cultivated, with no attempt at seed selection or improvement, and the plants are generally shorter and coarser than the hemp grown in Kentucky. The short oil-seed hemp with slender stems, about 30 inches high, bearing compact clusters of seeds and maturing in 60 to 90 days, is of little value for fiber production, but the experimental plats, grown from seed imported from Russia, indicate that it may be valuable as an oil-seed crop to be harvested and thrashed in the same manner as oil-seed flax.

HUNGARY

The hemp in Hungary has received more attention in recent years than that in Russia, and this has resulted in a better type of plants. An experimental plat grown at Washington from Hungarian seed attained a height of 6 to 10 feet in the seed row. The internodes were rather short, the branches numerous, curved upward, and bearing crowded seed clusters and small leaves. About one-third of the plants had dark-purple or copper-colored foliage and were more compact in habit than those with normal green foliage.

ITALY

The highest-priced hemp fiber in the markets of either America or Europe is produced in Italy, (Bruck, Werner F. Studien uber den Hanfbau in Italien, p. 7, 1911.) but it is obtained from plants similar to those in Kentucky. The higher price of the fiber is due not to superior plants, but to water retting and to increased care and labor in the preparation of the fiber. Four varieties are cultivated in Italy:

(1) "Bologna," or great hemp, called in France "chanvre de Piedmont," is grown in northern Italy in the provinces of Bologna, Ferrara, Roviga, and Modena. In the rich alluvial soils and under the intensive cultivation there practiced this variety averages nearly 12 feet in height, but it is said to

deteriorate rapidly when cultivated elsewhere. (2) "Cannapa picola," small hemp, attaining a height of 4 to 7 feet, with a rather slender reddish stalk, is cultivated in the valley of the Arno in the department of Tuscany. (Dodge, Charles Richards. Culture of hemp in Europe. U.S. Department of Agriculture, Fiber Investigations, Report No. 11, p. 5, 1898.) (3) "Neapolitan," large seeded. (4) "Neapolitan," small seeded.

The two varieties of Neapolitan hemp are cultivated in the vicinity of Naples, and even so far up on the sides of Vesuvius that fields of hemp are occasionally destroyed by the eruptions of that volcano. Seed of each of these Italian varieties has been grown in trial plats at Washington, D.C., and Lexington, Ky. The Bologna, or Piedmont, hemp in seed rows attained a height of 8 to 11 feet, nearly as tall as Kentucky seed hemp grown for comparison, but with thicker stalks, shorter and more rigid branches, and smaller and more densely clustered leaves. The small hemp, cannapa picola, was only 4 to 6 feet high. The large-seeded Neapolitan was 7 to 10 feet high, smaller than the Bologna, but otherwise more like Kentucky hemp, with more slender stalks and more open foliage. The small seeded Neapolitan, with seeds weighing less than 1 gram per 100, rarely exceeded 4 feet in height in the series of plats where all were tried.

FRANCE

Hemp is cultivated in France chiefly in the departments of Sarthe and Ille-et-Vilaine, in the valley of the Loire River. Two varieties are grown, the Piedmont, from Italian seed, and the common hemp of Europe. The former grows large and coarse, though not as tall as in the Bologna region, and it produces a rather coarse fiber suitable for coarse twines. The latter, seed of which is sown at the rate of 1 1/2 to 2 bushels per acre, has a very slender stalk, rarely more than 4 or 5 feet high, producing a fine flax-like fiber that is largely used in woven hemp linens. The common hemp of Europe, which includes the short hemp of France, is also cultivated to a limited extent in Spain, Belgium, and Germany. It grows taller and coarser when sown less thickly on rich land, but it never attains the size of the Bologna type.

CHILE

Chilean hemp, originally from seed of the common hemp of Europe, has developed in three and a half centuries into coarser plants with larger seeds. When sown broadcast for fiber in Chile the plants attain a height of 6 to 8 feet, and when in checks or drills for seed they reach 10 to 12 feet. Hemp from Chilean seed (S.P. I. No. 24307), grown at the experiment stations at Lexington, Ky., and St. Paul, Minn., in 1909, was 4 to 9 feet high in the broadcast plats and about the same height in the seed drills. It matured earlier than hemp of Chinese origin. Its leaves were small and crowded, with the seed clusters near the ends of slender, spreading branches. The fiber was coarse and harsh. The seeds were very large, 5 to 6 millimeters long, and weighed about 2 grams per 100.

TURKEY

A variety of hemp, intermediate between the fiber-producing and the typical drug-producing types, is cultivated in Asiatic Turkey, especially in the region of Damascus, and to a limited extent in European Turkey. This variety, called Smyrna, is about the poorest variety from which fiber is obtained. It is cultivated chiefly for the narcotic drug, but fiber is also obtained from the stalks. It grows 3 to 6 feet high, with short internodes, numerous ascending branches, densely crowded foliage of small leaves, and abundant seeds maturing early. It seems well suited for the production of birdseed, but its poor type, combined with prolific seed production, makes it a dangerous plant to grow in connection with fiber crops.

INDIA

Hemp is cultivated in India over an area of 2,000 to 5,000 acres annually for the production of the narcotic drugs known as hashish, charras, bhang, and ganja. Some fiber is obtained, especially from the staminate plants, in the northern part of Kashmir, where the hemp grown for the production of charras is more like the fiber types than that grown for bhang farther south. Plants grown by the Department of Agriculture at Washington from seed received from the Botanical Garden at Sibpur, Calcutta, India, agreed almost perfectly with the description of Cannabis indica (Lamarck. Encyclopedie, v. 1, p. 695, 1788.) written by Lamarck more than a century ago. (Pl. XLII, fig. 2.) They were distinctly different in general appearance from any of the numerous forms grown by this department from seed obtained in nearly all countries where hemp is cultivated, but the differences in botanical characters were less marked. The Indian hemp differed from Kentucky hemp in its more densely branching habit, its very dense foliage, the leaves mostly alternate, 7 to 11 (usually 9) very narrow leaflets, and in its nearly solid stalk. It was imperfectly dioecious, a character not observed in any other variety. Its foliage remained green until after the last leaves of even the pistillate plants of Kentucky hemp had withered and fallen. It was very attractive as an ornamental plant but of no value for fiber.

ARABIA AND AFRICA

Hemp is somewhat similar to that of India, but generally shorter, is cultivated in Arabia, northern Africa, and also by some of the natives in central and southern Africa for the production of the drug, but not for fiber. In Arabia it is called "takrousi," in Morocco "kief" or "kif," and in South Africa "dakkan." None of these plants is suitable for fiber production.

KENTUCKY

Practically all of the hemp grown in the United States is from seed produced in Kentucky. The first hemp grown in Kentucky was of European origin, the seed having been brought to the colonies, especially Virginia, and taken from there to Kentucky. In recent years there has been practically no importation of seed from Europe. Remnants of the European types are occasionally found in the shorter, more densely branching stalks terminating in thick clusters of small leaves. These plants yield more seed and mature earlier than the more desirable fiber types introduced from China. Nearly all of the hemp now grown in Kentucky is of Chinese origin. Small packets of seed are received from American missionaries in China. These seeds are carefully cultivated for two or three generations in order to secure a sufficient quantity for field cultivation, and also to acclimate the plants to Kentucky conditions. Attempts to produce fiber plants by sowing imported seed broadcast have not given satisfactory results. Seed of the second or third generation from China is generally regarded as most desirable. This Kentucky hemp of Chinese origin has long internodes, long, slender branches, opposite and nearly horizontal except the upper ones, large leaves usually drooping and not crowded, with the seeds in small clusters near the ends of the branches. Small, dark-colored seeds distinctly mottled are preferred by the Kentucky hemp growers. Under favorable conditions Kentucky hemp attains a height of 7 to 10 feet when grown broadcast for fiber and 9 to 14 feet when cultivated for seed.

IMPROVEMENT BY SEED INTRODUCTION

Without selection or continued efforts to maintain superior types, the hemp in Kentucky deteriorates. As stated by the growers, the hemp "runs out." The poorer types of plants for fiber are usually the most prolific seed bearers, and they are often earlier in maturing; therefore, without selection or roguing, the seed of these undesirable types increases more rapidly than that of the tall, late-maturing, better types which bear fewer seeds. New supplies of seed are brought from China to renew the stock. Owing to the confusion of names the seed received is not always of a desirable kind, and sometimes jute, China jute, or ramie seeds are obtained. When seed of the ta-ma variety is secured and is properly cultivated for two or three generations there is a marked improvement, but these improved strains run out in less than 10 years. The numerous trials that have been made by the Department of Agriculture with hemp seed from nearly all of the sources mentioned and repeated introductions from the more promising sources indicate that little permanent improvement may be expected from mere introduction not followed by breeding and continued selection. In no instance, so far as observed, have any of the plants from imported seed grown as well the first year as the Kentucky hemp cultivated for comparison. Further introduction of seed in small quantities is needed to furnish stock for breeding and selection. The most promising varieties for introduction are ta-ma and shan -ma-tze, from China; Hiroshima and Tochigi, from Japan; Bologna, from Italy; and improved types from Hungary.

IMPROVEMENT BY SELECTION

Kentucky hemp is reasonably uniform, not because of selection, or even grading the seeds, but because all types have become mixed together. Nearly all the seed is raised in a limited area. Hemp being cross-fertilized, it is more difficult to keep distinct types separate than in the case of wheat, flax, or other crops with self-pollinated flowers, but it is merely necessary to isolate the plants cultivated for seed and then exercise care to prevent the seed from becoming mixed. Until 1903 no well-planned and continued effort seems to have been undertaken in this country to produce an improved variety of hemp. At that time the results of breeding by careful selection improved varieties of wheat and flax at the Minnesota Agricultural Experiment Station were beginning to yield practical returns to the farmers of that State. Mr. Fritz Knorr, from Kentucky, then a student in the Minnesota College of Agriculture, was encouraged to take up the work with hemp. Seed purchased from a dealer in Nicholasville, Ky., was furnished by the United States Department of Agriculture. The work of selection was continued until 1909 under the direction of Prof. C. P. Bull, agronomist at the station. Points especially noted in selecting plants from which to save seed for propagation were length of internode, thinness of shell, height, and tendency of the stems to be well fluted. The seasons there were too short to permit selection for plants taking a longer season for growth. The improved strain of hemp thus developed was called Minnesota No. 8. Seed of this strain sown at the experiment station at Lexington, Ky., in 1910 and 1911 produced plants more uniform than those from unselected Kentucky seed, and the fiber was superior in both yield and quality. A small supply of this seed, grown by the Department of Agriculture at Washington, D. C., in 1912, was distributed to Kentucky hemp-seed growers in 1913, and in every instance the resulting seed plants were decidedly superior to those from ordinary Kentucky seed. Seed selection is practiced to a limited extent on some of the best hemp-seed farms in Kentucky. Before the seed-hemp plants are cut the grower goes through the field and marks the plants from which seed is to be saved for the seed crop of the following year. Plants are usually selected for height, lateness, and length of internodes. Continued selection in this manner will improve the type. Without selection continued each season, the general average of the crop deteriorates.

CLIMATE

Hemp requires a humid temperate climate, such as that throughout the greater part of the Mississippi Valley. It has been grown experimentally as far north as Saskatoon, in northwestern Canada, and as far south as New Orleans, La., and Brunswick, Ga.

TEMPERATURE

The best fiber-producing types of hemp require about four months free from killing frosts for the production of fiber and about five and one-half months for the full maturity of the seeds. The climatic conditions during the four months of the hemp-growing season in the region about Lexington, Ky., are indicated by the following table:

(Insert table from p. 305 here.) Table: Temperature and rainfall in the hemp-growing region of Kentucky (Henry, Alfred Judson. Climatology of the United States. U.S. Department of Agriculture, Weather Bureau, Bulletin Q, p.763, 1906.)

Hemp grows best where the temperature ranges between 60 degrees and 80 degrees F., but it will endure colder and warmer temperatures. Young seedlings and also mature plants will endure with little injury light frosts of short duration. Young hemp is less susceptible than oats to injury from frost, and fields of hemp ready for harvest have been uninjured by frosts which ruined fields of corn all around them. Frosts are injurious to nearly mature plants cultivated for seed production.

RAINFALL

Hemp requires a plentiful supply of moisture throughout its growing season, and especially during the first six weeks. After it has become well rooted and the stalks are 20 to 30 inches high it will endure drier conditions, but a severe drought hastens its maturity and tends to dwarf its growth. It will endure heavy rains, or even a flood of short duration, on light, well-drained soils, but on heavy, impervious soils excessive rain, especially when the plants are young will ruin the crop. In 1903, a large field of hemp on rich, sandy-loam soil of alluvial deposit, well supplied with humus, near Gridley, Cal., was flooded to a depth of 2 to 6 inches by high water in the Feather River. The hemp had germinated but a few days before and was only 1 to 3 inches high. The water remained on the land about three days. The hemp started slowly after the water receded, but in spite of the fact that there was no rain from this time, the last of March, until harvest, the last of August, it made a very satisfactory crop, 6 to 12 feet in height. The soil, of porous, spongy texture, remained moist below the dusty surface during the entire growing season.

An experimental crop of about 15 acres on impervious clay and silt of alluvial deposit, but lacking in humus, in eastern Louisiana was completely ruined by a heavy rain while the plants were small. The total average rainfall during the four months of the hemp-growing season in Kentucky is 15.6 inches, as shown in the table on page 305, and this is distributed throughout the season. When there is an unusual drought in that region, as in 1913, the hemp is severely injured. It is not likely to succeed on upland soils in localities where corn leaves curl because of drought before the middle of August.

IRRIGATION

In 1912, and again in 1913, crops of hemp were cultivated under irrigation at Lerdo, Cal. The soil there is an alluvial sandy loam of rather firm texture, but with good natural drainage and not enough clay to form a crust on the surface after flooding with water. The land is plowed deeply, leveled, and

197

made up into irrigation blocks with low borders over which drills and harvesting machinery may easily work. The seed is drilled in the direction of the fall, so that when flooded the water runs slowly down the drill furrows. Three irrigations are sufficient, provided the seed is sown early enough to get the benefit of the March rains. The fiber thus produced is strong and of good quality.

WEATHER FOR RETTING AND BREAKING

Cool, moist weather, light snows or alternate freezing and thawing are favorable for retting hemp. Dry weather, not necessarily free from rain but with a rather low relative humidity, is essential for satisfactory work in breaking hemp. The relative humidity at Lexington in January, February, and March, when most of the hemp is broken, ranges from 62 to 82 per cent. The work of breaking hemp is rarely carried on when there is snow on the ground. The work of collecting and cleaning hemp seed can be done only in dry weather.

SOIL

SOILS IN THE HEMP-GROWING REGION OF KENTUCKY

The soil in most of the hemp fields of Kentucky is of a yellowish clay loam, often very dark as a result of decaying vegetable matter, and most of it overlying either Lexington or Cincinnati limestone. There are frequent outcroppings of lime rock throughout the region. The soil is deep, fertile, well supplied with humus, and its mechanical condition is such that it does not quickly dry out or become baked and hard. The land is rolling, affording good natural drainage.

HEMP SOILS IN OTHER STATES

In eastern Nebraska, hemp has been grown on a deep clay-loam prairie soil underlain with lime rock. In some of the fields there are small areas of gumbo soil, but hemp does not grow well on these areas. In California, hemp is cultivated on the reclaimed lands of alluvial deposits in the lower valley of the Sacramento River. This is a deep soil made up of silt and sand and with a very large proportion of decaying vegetable matter. These rich, alluvial soils, which are never subject to drought, produce a heavier growth of hemp than the more shallow upland soils in Kentucky. In Indiana, crops of hemp have been grown in the Kankakee Valley on peaty soils overlying marl or yellow clay containing an abundance of lime. These lands have been drained by large, open ditches. There is such a large proportion of peat in the soil that it will burn for months if set on fire during the dry season, yet this soil contains so much lime that when the vegetation is cleared away Kentucky bluegrass comes in rather than sedges. It is an alkaline rather than an acid soil. The large amount of peat gives these soils a loose, spongy texture, well adapted to hold moisture during dry seasons. Water remains in the ditches 6 to 10 feet below the surface nearly all summer, and the hemp crops have not been affected by the severe drought which has injured other crops on the surrounding uplands. In southeastern Pennsylvania, and in Indiana, Wisconsin, and Minnesota, the best crops, producing the largest yields of fiber and fiber of the best quality, have been grown on clay-loam upland soils. In some instances, however, the upland crops have suffered from drought.

SOILS SUITED TO HEMP

Hemp requires for the best development of the plant, and also for the production of a large quantity and good quality of fiber, a rich, moist soil having good natural drainage, yet not subject to severe drought at any time during the growing season. A clay loam of rather loose texture and containing a plentiful supply of decaying vegetable matter or an alluvial deposit alkaline and not acid in reaction should be chosen for this crop.

SOILS TO BE AVOIDED

Hemp will not grow well on stiff, impervious, clay soils, or on light sandy or gravelly soils. It will not grow well on soils that in their wild state are overgrown with either sedges or huckleberry bushes. These plants usually indicate acid soils. It will make only a poor growth on soils with a hardpan near the surface or in fields worn out by long cultivation. Clay loams or heavier soils give heavier yields of strong but coarser fiber than are obtained on sandy loams and lighter soils.

EFFECT OF HEMP ON THE LAND

Hemp cultivated for the production of fiber, cut before the seeds are formed and retted on the land where it has been grown, tends to improve rather than injure the soil. It improves its physical condition, destroys weeds, and does not exhaust its fertility.

PHYSICAL CONDITION

Hemp loosens the soil and makes it more mellow. The soil is shaded by hemp more than by any other crop. The foliage at the top of the growing plants makes a dense shade and, in addition, all of the leaves below the top fall off, forming a mulch on the ground, so that the surface of the soil remains moist and in better condition for the action of soil bacteria. The rather coarse taproots (Pl. XLI, fig. 3), penetrating deeply and bringing up plant food from the subsoil, decay quickly after the crop is harvested and tend to loosen the soil more than do the fibrous roots of wheat, oats, and similar broadcast crops. Land is more easily plowed after hemp than after corn or small grain.

HEMP DESTROYS WEEDS

Very few of the common weeds troublesome on the farm can survive the dense shade of a good crop of hemp. If the hemp makes a short, weak growth, owing to unsuitable soil, drought, or other causes, it will have little effect in checking the growth of weeds, but a good, dense crop, 6 feet or more in height,

HEMP GROWS UP TO 4 CROPS A YEAR AND ENRICHES THE SOIL'S FERTILITY

will leave the ground practically free from weeds at harvest time. In Wisconsin, Canada thistle has been completely killed and quack-grass severely checked by one crop of hemp. In one 4-acre field in Vernon County, Wis., where Canada thistles were very thick, fully 95 per cent of the thistles were killed where the hemp attained a height of 5 feet or more, but on a dry, gravelly hillside in this same field where it grew only 2 to 3 feet high, the thistles were checked no more than they would have been in a grain crop. Some vines, like the wild morning-glory and bindweed climb up the hemp stalks and secure light enough for growth, but low growing weeds can not live in a hemp field.

An abundant supply of plant food is required by hemp, but most of it is merely borrowed during development and returned to the soil at the close of the season. The amounts of the principal fertilizing elements contained in mature crops of hemp, as compared with other crops, are shown in the accompanying table.

Amounts of principal fertilizing elements in an acre of hemp, corn, wheat, oats, sugar beets, and cotton. (Insert first table from p. 310 here)

The data in the table indicate that hemp requires for its best development a richer soil than any of the other crops mentioned except sugar beets. These other crops, except the stalks of corn and the tops of beets, are entirely removed from the land, thus taking away nearly all the plant food consumed in their growth. Only the fiber of hemp is taken away from the farm and this is mostly cellulose, composed of water and carbonic acid. The relative proportions by weight of the different parts of the hemp plant, thoroughly air dried, are approximately as follows: Roots 10 per cent, stems 60 per cent, and leaves 30 per cent. The mineral ingredients of these different parts of the hemp plant are shown in the following table:

(Insert second table from p. 310 here. Peter, Robert. Chemical Examination of the Ash of Hemp and Buckwheat Plants. Kentucky Geological Survey, p. 12, 1884.)

The foliage, constituting nearly one-third of the weight of the entire plant and much richer in essential fertilizing elements than the stalks, all returns to the field where the hemp grows. The roots also remain and together with the stubble, they constitute more than 10 per cent of the total weight and contain approximately the same proportions of fertilizing elements as the stalks. The leaves and roots therefore return to the soil nearly two-thirds of the fertilizing elements used in building up the plant. After the hemp is harvested it is spread out on the same land for retting. In this retting process nearly all of the soluble ingredients are washed out and returned to the soil. When broken in the field on small hand brakes, as is still the common practice in Kentucky, the hurds, or central woody portion of the stalk, together with most of the outer bark, are left in small piles and burned, returning the mineral ingredients to the soil. Where machine brakes are used the hurds may serve an excellent purpose as an absorbent in stock yards and pig pens, to be returned to the fields in barnyard manure.

The mineral ingredients permanently removed from the farm are thus reduced to the small proportions contained in the fiber. These proportions, calculated in pounds per acre and compared with the amounts removed by other crops, are shown in the following table:

Mineral ingredients removed from the soil by hemp, wheat, corn, and tobacco, calculated in pounds per acre. (Peter, Robert. Chemical Examination of the Ash of Hemp and Buckwheat Plants. Kentucky Geological Survey, p. 17, 1884.) (Insert table from p. 311 here.)

The hemp fiber analyzed was in the ordinary condition as it leaves the farm. When washed with cold water, removing some but not all of the dirt, the ashy residue was reduced more than one-third, and the total earthy phosphates were reduced nearly one-half. The amount of plant food actually removed from the soil by hemp is so small as to demand little attention in considering soil exhaustion. The depletion of the humus is the most important factor, but even in this respect hemp is easier on the land than other crops except clover and alfalfa. The fact that hemp is often grown year after year on the same land for 10 to 20 years, with little or no application of fertilizer and very little diminution in yield, is evidence that it does not exhaust the soil.

ROTATION OF CROPS

In Kentucky, hemp is commonly grown year after year on the same land without rotation. It is the common practice in that State to sow hemp after bluegrass on land that has been in pasture for many years, or sometimes it is sown as the first crop on recently cleared timberland. It is then sown year after year until it ceases to be profitable or until conditions favor the introduction of other crops. On the prairie soils in eastern Nebraska and also on the peaty soils in northern Indiana, more uniform crops were obtained after the first year. On some of the farms in California hemp is grown in rotation with beans. Hemp is recommended to be grown in rotation with other farm crops on ordinary upland soils suited to its growth. In ordinary crop rotations it would take about the same place as oats. If retted on the same land, however, it would occupy the field during the entire growing season, so that it would be impossible to sow a field crop after hemp unless it were a crop of rye. The growing of rye after hemp has been recommended in order to prevent washing and to retain the soluble fertilizing elements that might otherwise be leached out during the winter. This recommendation, however, has not been put in practice sufficiently to demonstrate that it is of any real value. Hemp will grow well in a fertile soil after any crop, and it leaves the land in good condition for any succeeding crop. Hemp requires a plentiful supply of fertilizing elements, especially nitrogen, and it is therefore best to have it succeed clover, peas, or grass sod. If it follows wheat, oats, or corn, these crops should be well fertilized with barnyard manure. The following crop rotations are suggested for hemp on fertile upland soils: (Insert chart from p. 313 here).

Hemp leaves the ground mellow and free from weeds and is therefore recommended to precede sugar beets, onions, celery, and similar crops which require hand weeding. If hemp is grown primarily to kill Canada thistle, quack-grass, or similar perennial weeds, it may be grown repeatedly on the same land until the weeds are subdued.

FERTILIZERS

Hemp requires an abundant supply of plant food. Attaining in four months a height of 6 to 12 feet and producing a larger amount of dry vegetable matter than any other crop in temperate climates, it must be grown on a soil naturally fertile or enriched by a liberal application of fertilizer. In Europe and in Asia heavy applications of fertilizers are used to keep the soils up to the standard for growing hemp, but in the United States most of the hemp is grown on lands the fertility of which has not been exhausted by centuries of cultivation. In Kentucky, where the farms are well stocked with horses and cattle, barnyard manure is used to maintain the fertility of the soils, but it is usually applied to other crops and not directly to hemp. In other States no fertilizer has been applied to soils where hemp is grown, except in somewhat limited experiments.

BARNYARD MANURE.---The best single fertilizer for hemp is undoubtedly barnyard manure. It supplies the three important plant foods, nitrogen, potash, and phosphoric acid, and it also adds to the store of humus, which appears to be more necessary for hemp than for most other farm crops. If other fertilizers are used, it is well to apply barnyard manure also, but it should be applied to the preceding crop, or, at the latest, in the fall before the hemp is sown. It must be well rotted and thoroughly mixed with the soil before the hemp seed is sown, so as to promote a uniform growth of the hemp stalks. Uniformity in the size of the plants of other crops is of little consequence, but in hemp it is a matter of prime importance. An application of coarse manure in the spring, just before sowing, is likely to result in more injury than benefit. The amount that may be applied profitably will vary with different soils. There is little danger, however, of inducing too rank a growth of hemp on upland soils, provided the plants are uniform, for it must be borne in mind that stalk and not fruit is desired. On soils deficient in humus as the result of long cultivation, the increased growth of hemp may well repay for the application of 15 to 20 tons of barnyard manure per acre. It would be unwise to sow hemp on such soils until they had been heavily fertilized with barnyard manure.

COMMERCIAL FERTILIZERS.---On worn-out soils, peaty soils, and possibly on some alluvial soils, commercial fertilizers may be used with profit in addition to barnyard manure. The primary effect to be desired from commercial fertilizers on hemp is a more rapid growth of the crop early in the season. This rapid early growth usually results in a greater yield and better quality of fiber. The results of a series of experiments conducted at the agricultural experiment station at Lexington, Ky., in 1889 led to the following conclusions: (Scovel, M.A. Effect of Commercial Fertilizers on Hemp. Kentucky Agricultural Experiment Station, Bulletin 27, p. 3, 1890.) (1) That hemp can be raised successfully on worn bluegrass soils with the aid of commercial fertilizers. (2) That both potash and nitrogen are required to produce the best results. (3) That the effect was the same, whether muriate or sulphate was used to furnish potash. (4) That the effect was about the same, whether nitrate of soda or sulphate of ammonia was used to furnish nitrogen. (5) That a commercial fertilizer containing about 6 per cent of available phosphoric acid, 12 per cent of actual potash, and 4 per cent of nitrogen (mostly in the form of nitrate of soda or sulphate of ammonia) would be a good fertilizer for trial. The increased yield and improved quality of the fiber on the fertilized plats compared with the yield from the check plat, not fertilized, in these experiments would warrant the application of nitrogen at the rate of 160 pounds of nitrate of soda or 120 pounds of sulphate of ammonia per acre, and potash at the rate of about 160 pounds of either sulphate or muriate of potash per acre. On the rich alluvial soils reclaimed by dikes from the Sacramento River at Courtland, Cal., Mr. John Heaney has found that an application of nitrate of soda at the rate of not more than 100 pounds per acre soon after sowing and again two weeks to a month later, or after the first application has been washed down by rains, will increase the yield and improve the quality of the fiber.

LEGUMINOUS CROPS OR GREEN MANURE.---Beans grown before hemp and the vines returned to the land and plowed under have given good results in increased yield and improved quality of fiber on alluvial soils at Courtland, Cal. Clover is sometimes plowed under in Kentucky to enrich the land for hemp. It must be plowed under during the preceding fall, so as to become thoroughly rotted before the hemp is grown.

HEMP AS A GREEN MANURE.---In experiments with various crops for green manure for wheat in India, hemp was found to give the best results. (Report of Cawnpore Agricultural Station, United Provinces, India, for 1908, p. 12.) In exceptionally dry seasons, as in 1908 and 1913, many fields of hemp do not grow high enough to be utilized profitably for fiber production. They are *often left until fully mature and then burned. Better results would doubtless be obtained if the hemp were plowed under as soon as it could be determined that it would not make a sufficient growth for fiber production. Mature hemp stalks or dry hurds should not be plowed under, because they rot very slowly.*

DISEASES, INSECTS, AND WEEDS

Hemp is remarkably free from diseases caused by fungi. In one instance at Havelock, Nebr., in a low spot where water had stood, nearly 3 per cent of the hemp plants were dead. The roots of these dead plants were pink in color and a fungous mycelium was found in them, but it was not in a stage of development to permit identification. The fungus was probably not the primary cause of the trouble, since the dead plants were confined to the low place and there was no recurrence of the disease on hemp grown in the same field the following year.

A fungus described under the name Dendrophoma marconii Cav. was observed on hemp in northern Italy in 1887. (Cavara, Fridiano. Appunti di Patologia Vegetal. Atti dell' Instituto Botanico dell' Universita di Pavia, s. 2, v. 1, p. 425, 1888.) This fungus attacked the plants after they were mature enough to harvest for fiber. Its progress over the plant attacked and also the distribution of the infection over the field were described as very rapid, but if the disease is discovered at its inception and the crop promptly harvested it causes very little damage.

In the fall of 1913 a disease was observed on seed hemp grown by the Department of Agriculture at Washington. (Pl. XLIII, fig. 2.) It did not appear until after the stage of full flowering of the staminate plants and therefore after the stage for harvesting for fiber. A severe hailstorm had bruised the plants and broken the bark, doubtless making them more susceptible to the disease. The first symptoms noted in each plant attacked were wilted leaves near the ends of branches above the middle of the plant, accompanied by an area of discolored bark on the main stalk below the base of each diseased branch. In warm, moist weather the disease spread rapidly, killing a plant 10 feet high in five days and also infesting other plants. It was observed only on pistillate plants, but the last late-maturing staminate plants left in the plat after thinning the earlier ones were cut soon after the disease was discovered. (This fungus was not in a stage permitting identification, but cultures for further study were made in the Laboratory of Plant Pathology.)

In a few instances insects boring in the stems have killed some plants, but the injury caused in this manner is too small to be regarded as really troublesome.

Cutworms have caused some damage in the late-sown hemp in land plowed in the spring, but there is practically no danger from this source in hemp sown at the proper season and in fall-plowed land well harrowed before sowing.

A Chilean dodder (Cuscuta racemosa) troublesome on alfalfa in northern California was found on the hemp at Gridley, Cal., in 1903. Although it was abundant in some parts of the field at about the time the hemp was ready for harvest, it did not cause any serious injury. Black bindweed (Polygonum convolvulus) and wild morning-glory (Convolvulus sepium) sometimes cause trouble in low, rich land by climbing up the plants and binding them together.

The only really serious enemy to hemp is branched broom rape (Orobanche ramosa). (Pl. XLIII, fig. 3.) This is a weed 6 to 15 inches high, with small, brownish yellow, scalelike leaves and rather dull purple flowers. The entire plant is covered with sticky glands which catch the dust and give it a dirty appearance. Its roots are parasitic on the roots of hemp. It is also parasitic on tobacco and tomato roots. (Garman, H. The Broom-Rape of Hemp and Tobacco. Kentucky Agricultural Experiment Station, Bulletin 24, p. 16, 1890.) Branched broom rape is troublesome in Europe and the United States, but is not known in Asia. Its seeds are very small, about the size of tobacco seed, and they stick to the gummy calyx surrounding the hemp seed when the seed-hemp plants are permitted to fall on the ground in harvesting. There is still more opportunity for them to come in contact with the seed of hemp grown for fiber. The broom rape is doubtless distributed more by means of lint seed (seed from overripe fiber hemp) than by any other means. When broom rape becomes abundant it often kills a large proportion of the hemp plants before they reach maturity. As a precaution it is well to sow only well-cleaned seed from cultivated hemp and insist on a guaranty of no lint seed. If the land becomes infested, crops other than hemp, tobacco, tomatoes, or potatoes should be grown for a period of at least seven years. The seeds retain their vitality several years. (Garman, H. The Broom-Rapes. Kentucky Agricultural Experiment Station, Bulletin 105, p. 14, 1903.)

HEMP-SEED PRODUCTION

All of the hemp seed used in the United States for the production of hemp for fiber is produced in Kentucky. Nearly all of it is obtained from plants cultivated especially for seed production and not for fiber. The plants cultivated for seed for the fiber crop are of the fiber-producing type and not the type commonly obtained in bird-seed hemp. Old stocks of hemp seed of low vitality are often sold for bird seed, but much of the hemp seed sold by seedsmen or dealers in bird supplies is of the densely branching Smyrna type.

LINT SEED

In some instances seed is saved from hemp grown for fiber but permitted to get overripe before cutting. This is known as lint seed. It is generally regarded as inferior to seed from cultivated plants. A good crop is sometimes obtained from lint seed, but it is often lacking in vigor as well as germinative vitality, and it is rare that good crops are obtained from lint seed of the second or third generation.

CULTIVATED SEED

Nearly all of the cultivated seed is grown in the valley of the Kentucky River and along the creeks tributary to this river for a distance of about 50 miles above High Bridge. The river through this region flows in a deep gorge about 150 feet below the general level of the land. The sides of this valley are steep, with limestone outcropping, and in some places perpendicular ledges of lime rock in level strata. (Pl. XLII, fig. 3.) The river, which overflows every spring, almost covering the valley between the rocky walls, forms alluvial deposits from a few rods to half a mile in width. The seed hemp is grown on these inundated areas, and especially along the creeks, where the water from the river backs up, leaving a richer deposit of silt than along the banks of the river proper, where the deposited soils are more sandy. There is a longer season free from frost in these deep valleys than on the adjacent highlands. Instead of having earlier frosts in the fall, as may be usually expected in lowlands, the valley is filled with fog on still nights, thus preventing damage from frost. For the production of hemp seed a rich, alluvial soil containing a plentiful supply of lime and also a plentiful supply of moisture throughout the growing season is necessary. The crop also requires a long season for development. The young seedlings will endure light frosts without injury, but a frost before harvest will nearly ruin the crop. A period of dry weather is necessary after the harvest in order to beat out and clean the seeds.

PREPARATION OF LAND

The land is plowed as soon as possible after the spring floods, which usually occur in February and early in March. After harrowing, it is marked in checks about 4 or 5 feet each way. Hemp cultivated for seed production must have room to develop branches. (Pl. XL, fig. 1.)

PLANTING

The seed is planted between the 20th of March and the last of April--- usually earlier than the seed is sown for the production of fiber. It is usually planted by hand, 5 to 7 seeds in a hill, and covered with a hoe. In some instances planters are used, somewhat like those used for planting corn, and on some farms seeders are used which plant 1 or 2 drills at a time 4 or 5 feet apart. When planted in drills it is usually necessary to thin out the plants afterwards. One or two quarts of seed are sufficient to plant an acre. Less than one quart would be sufficient if all the plants were allowed to grow.

CULTIVATION

On the best farms the crop is cultivated four times---twice rather deep and twice with cultivators with fine teeth, merely stirring the surface. When the first flowers are produced, so that the staminate plants may be recognized, all of these plants are cut out except about one per square rod. These will produce sufficient pollen to fertilize the flowers on the pistillate, or seed-bearing plants, and the removal of the others will give more room for the development of the seed-bearing plants.

HARVEST

The seed-bearing plants are allowed to remain until fully mature, or as long as possible without injury from frost. They are cut with corn knives, usually during the first half of October, leaving the stubble 10 to 20 inches high. The plants are set up in loose shocks around one or two plants which have been left standing. The shocks are usually bound near the top with binder twine. They are left in this manner for two or three weeks, until thoroughly dry. (Pl. XLIII, fig. 1.)

COLLECTING THE SEED

When the seed hemp is thoroughly dry, men (usually in gangs of five or six, with tarpaulins about 20 feet square) go into the field. One man with an ax cuts off the hemp stubble between four shocks and clears a space large enough to spread the tarpaulin. The other men pick up an entire shock and throw it on the tarpaulin. They then beat off the seeds with sticks about 5 feet long and 1 1/2 inches in diameter. (Pl. XLIV. fig. 1.) When the seed has been beaten off from one side of the shock the men turn it over by means of the sticks, and after beating off all of the seed they pick up with the sticks the stalks in one bunch and throw them off the canvas, and then treat another shock in the same manner. They will beat off the seed from four shocks in 15 to 20 minutes, securing 2 or 3 pecks of seed from each shock. While this seems a rather crude way of collecting the seed, it is doubtless the most economical and practical method that may be devised. The seed falls so readily from the dry hemp stalks that it would be impossible to move them without a very great loss. Furthermore, it would be very difficult to handle plants 10 to 14 feet high, with rigid branches 3 to 6 feet in length, so as to feed them to any kind of thrashing machine.

CLEANING THE SEED

The seed and chaff which have been beaten on the tarpaulin are sometimes beaten or tramped to break up the coarser bunches and stalks, and in some instances they are rubbed through coarse sieves in order to reduce them enough to be put through a fanning mill. The seed is then partly cleaned by a fanning mill in the field and afterwards run once or twice through another mill with finer sieves and better adjustments of fans. Even after this treatment it is usually put through a seed-cleaning machine by the dealers. There has recently been introduced on some of the best seed-hemp farms a kind of homemade *thrashing machine, consisting essentially of a feeding device, cylinder, and concaves, attached to a rather large fanning mill, all being driven by a gasoline engine. (Pl. XLIV, fig. 2.) The hemp seed is fed to this machine just as it comes from the tarpaulin after beating off from the shock. It combines the process of breaking up the chaff into finer pieces and the work of fanning the seed in the field, and it performs this work more effectively and more rapidly.*

YIELD

Under favorable conditions the yield of hemp seed ranges from 12 to 25 bushels per acre. From 16 to 18 bushels are regarded as a fair average yield.

COST OF SEED PRODUCTION

The hemp-seed growers state that it costs about $2.50 per bushel to produce hemp seed, counting the annual rental of the land at about $10 per acre. With the introduction of improved machinery for cleaning the hemp this cost may be somewhat reduced, since it is estimated that with the ordinary methods of rubbing the seed through sieves or beating it to reduce the chaff to finer pieces the cost from beating it off the shock to delivering it at the market is about 50 cents per bushel. These estimates of cost are based on wages at $1.25 per day.

PRICES

The price of hemp seed, as sold by the farmer during the past 10 years, has ranged from $2.50 to $5 per bushel. The average farm price during this period has been not far from $3 per bushel. Hemp seed is sold by weight, a bushel weighing 44 pounds.

CULTIVATION FOR FIBER

PREPARATION OF THE LAND

Fall plowing on most soils is generally regarded as best for hemp, since the action of the frost in winter helps to disintegrate the particles of soil, making it more uniform in character. In practice, hemp land is plowed at any time from October to late seeding time in May, but hemp should never be sown on spring-plowed sod. The land should be plowed 8 or 9 inches in order to give a deep seed bed and opportunity for root development. Plowing either around the field or from the center is recommended, since back furrows and dead furrows will result in uneven moisture conditions and more uneven hemp. Before sowing, the land is harrowed to make a mellow seed bed and uniform level surface. Sometimes this harrowing is omitted, especially when hemp is grown on stubble ground plowed just before seeding. Harrowing or leveling in some manner is recommended at all times, in order to secure conditions for covering the seed at a uniform depth and also to facilitate close cutting at harvest time.

SEEDING

METHODS OF SEEDING

Hemp seed should be sown as uniformly as possible all over the ground and covered as nearly as possible at a uniform depth of about three-fourths of an inch, or as deep as 2 inches in light soils. Ordinary grain drills usually plant the seed too deeply and in drills too far apart for the best results. Uniform distribution is sometimes secured by drilling in both directions. This double working, especially with a disk drill, leaves the land in good condition. Ordinary grain drills do not have a feed indicator for hemp seed, but they may be readily calibrated, and this should be done before running the risk of sowing too much or too little. Fill the seed box with hemp seed, spread a canvas under the feeding tubes, set the indicator at a little less than one-half bushel per acre for wheat, and turn the drivewheel as many times as it would turn in sowing one-tenth acre. One method giving good results is to remove the lower sections of the feeding tubes on grain drills and place a flat board so that the hemp seed falling against it will be more evenly distributed, the seed being covered either by the shoes of the drill or by a light barrow. Good results are obtained with disk drills, roller press drills, and also with the end-gate broadcast seeder. Drills made especially for sowing hemp seed are now on the market, and they are superseding all other methods of sowing hemp seed in Kentucky. Rolling after seeding is advised, in order to pack the soil about the seed and to secure a smooth surface for cutting, but rolling is not recommended for soils where it is known to have an injurious effect.

AMOUNT OF SEED

Hemp is sown at the rate of about 3 pecks (33 pounds) per acre. On especially rich soil 1 1/2 bushels may be sown with good results, and on poor land that will not support a dense, heavy crop a smaller amount is recommended. If conditions are favorable and the seed germinates 98 to 100 per cent, 3 pecks are usually sufficient. When kept dry, hemp seed retains its germinative vitality well for at least three or four years, but different lots have been found to vary from 35 to 100 per cent, and it is always well to test the seed before sowing.

TIME OF SEEDING

In Kentucky, hemp seed is sown from the last of March to the last of May. The best results are usually obtained from April seeding. Later seedings may be successful when there is a plentiful rainfall in June. In Nebraska, hemp seed was sown in April, May, or sometimes as late as June. In California it is sown in February or March; in Indiana and Wisconsin, in May. In general, the best time for sowing hemp seed is just before the time for sowing oats in any given locality. After the seed is sown, the hemp crop requires no further care or attention until the time of harvest.

HARVEST

TIME

In California, hemp is cut late in July or in August; in Kentucky, Indiana, and Wisconsin it is cut in September. The hemp should be cut when the staminate plants are in full flower and the pollen is flying. If cut earlier, the fiber will be finer and softer but also weaker and less in quantity. If permitted to become overripe, the fiber will be coarse, harsh, and less pliable, and it will be impossible to ret the stalks properly.

METHODS OF HARVESTING

HARVESTING BY HAND

In Kentucky, a small portion of the hemp crop is still cut by hand with a reaping knife or hemp hook. (Pl. XLV, fig. 1.) This knife is somewhat similar to a long-handled corn cutter. The man cutting the hemp pulls an armful of stalks toward him with his left arm and cuts them off as near the base as possible by drawing the knife close to the ground; he then lays the stalks on the ground in a smooth, even row, with the butts toward him, that is, toward the uncut hemp. An experienced hand will cut with a reaping knife about three-fourths of an acre a day. The hemp stalks are allowed to lie on the ground until dry, when they are raked up by hand and set up in shocks until time to spread for retting.

HARVESTING WITH REAPERS

Sweep-rake reapers are being used in increasing numbers for harvesting hemp in Kentucky and in all other localities where hemp is raised. (Pl. XLV, fig. 2.) While not entirely satisfactory, they are being improved and strengthened so as to be better adapted for heavy work. Three men, one to grind sections, one to drive, and one to attend to the machine, and four strong horses or mules are required in cutting hemp with a reaper. Under favorable conditions, from 5 to 7 acres per day can be cut in this manner. This more rapid work makes it possible to harvest the crop more nearly at the proper time. The stalks, after curing in the gavel, are set up in shocks, usually without binding into bundles unless they are to be stacked.

HARVESTING WITH MOWING MACHINES

In some places hemp is cut with ordinary mowing machines. (Pl. XLV, fig. 3.) A horizontal bar nearly parallel with the cutting bar, the outer end projecting slightly forward, is attached to an upright fastened to the tongue of the machine. This bar is about 4 feet above the cutting bar and about 20 inches to the front. It bends the hemp stalks over in the direction the machine is going. The stalks are more easily cut when thus bent away from the knives and, furthermore, the bases snap back of the cutting bar and never drop through between the guards to be cut a second time, as they often do when cut standing erect. With a 5 1/2- foot mowing machine thus equipped, one man and one team of two horses will cut 6 to 8 acres per day. The work is regarded as about equal to cutting a heavy crop of clover. The hemp thus cut all falls in the direction the machine is going, the tops overlapping the butts of the stalks. The ordinary track clearer at the end of the bar clears a path, so that the stalks are not materially injured either by the horses or the wheels of the machine at the next round.

The hemp stalks are then left where they fall until retted, or in places where the crop is heavy the stalks are turned once or twice to secure uniform curing and retting. When sufficiently retted the stalks are raked up with a 2-horse hayrake, going crosswise of the swaths, and then drawn, like hay, to the machine brake. This is the most inexpensive method for handling the crop. It is impossible to make clean, long, straight fiber from stalks handled in this manner, and it is not recommended where better methods are practicable. It is worthy of more extended use, however, for handling short and irregular hemp, and hundreds of acres of hemp now burned in Kentucky because it is too short to be treated in the regular manner might be handled with profit by this method. There may be nearly as much profit in 3 1/2-cent fiber produced at a cost of 2 cents per pound as in 5-cent fiber produced at a cost of 3 cents, provided the land rent is not too large an item of cost.

NEED FOR IMPROVEMENT IN HEMP HARVESTERS

The most satisfactory hemp-harvesting machines now in use are the self-rake reapers, made especially for this purpose. They are just about as satisfactory for hemp now as the similar machines for wheat and oats were 30 years ago. More efficient harvesting machinery is needed to bring the handling of this crop up to present methods in harvesting corn or small grain. A machine is needed which will cut the stalks close to the ground, deliver them straight and not bruised or broken, with the butts even, and bound in bundles about 8 inches in diameter. A modified form of the upright corn binder, arranged to cut a swath about 4 feet wide, is suggested. Modified forms of grain binders have been tried, but with rather unsatisfactory results. Green hemp 8 to 14 feet high can not be handled successfully by grain binders; furthermore, the reel breaks or damages a large proportion of the hemp. The tough, fibrous stalks, some of which may be an inch in diameter, are more difficult to cut than grain and therefore require sharp knives with a high motion. A hemp-reaping machine is also needed that will cut the hemp and lay it down in an even swath, as grain is laid with a cradle. The butts should all be in one direction, and the swath should be far enough from the cut hemp so as not to be in the way at the next round. A machine of this type may be used where it is desired to ret the hemp in the fall immediately after cutting. It might be used for late crops in Kentucky, or generally for hemp farther north, where there is little danger of "sunburn" after the hemp is harvested.

STACKING

Hemp stalks which are to be stacked are bound in bundles about 10 inches in diameter, with small hemp plants for bands, before being placed in shocks. (Pl. XLVI, fig. 2.) They are allowed to stand in the shock from 10 to 15 days, or a sufficient length of time to avoid danger of heating in the stack. The bundles are hauled from the shocks to the stacks in rather small loads of half a ton or less on a low rack or sled. Three men with a team and low wagon to haul the stalks can put up two hemp stacks of about 8 tons each in a day.

A hemp stack must be built to shed water. It is started much like a grain stack with a shock, around which the bundles are placed in tiers, with the butts sloping downward and outward. The stack is kept higher in the center and each succeeding outer tier projects slightly to a height of 5 or 6 feet, when another shock is built in the center, around which the bundles are carefully placed to shed water and the peak capped with an upright bundle. A well-built stack may be kept four or five years without injury. Hemp which has been stacked rets more quickly and more evenly, the fiber is usually of better quality, and the yield of fiber is usually greater than from hemp retted directly from the shock. Hemp is stacked before retting, but not after retting in Kentucky. Stacking retted hemp stalks for storage before breaking is not recommended in climates where there is danger of gathering moisture. Retted stalks may be stored in sheds where they will be kept dry.

CARE IN HANDLING

Hemp stalks must be kept straight, unbroken, and with the butts even. They must be handled with greater care than is commonly exercised in handling grain crops. When a bunch of loose stalks is picked up at any stage of the operation, it is chucked down on the butts to make them even. The loose stalks, or bundles, are handled by hand and not with pitchforks. The only tool used in handling the stalks is a hook or rake, in gathering them up from the swath.

RETTING

Retting is a process in which the gums surrounding the fibers and binding them together are partly dissolved and removed. It permits the fiber to be separated from the woody inner portion of the stalk and from the thin outer bark, and it also removes soluble materials which would cause rapid decomposition if left with the fiber. Two methods of retting are practiced commercially, viz, dew retting and water retting.

DEW RETTING

In this country dew retting is practiced almost exclusively. The hemp is spread on the ground in thin, even rows, so that it will all be uniformly exposed to the weather. In spreading hemp the workman takes an armful of stalks and, walking backward, slides them sidewise from his knee, so that the butts are all even in one direction and the layer is not more than three stalks in thickness.

(Pl. XLIV, fig. 3.) This work is usually paid for at the rate of $1 per acre, and experienced hands will average more than 1 acre per day. The hemp is left on the ground from four weeks to four months. Warm, moist weather promotes the retting process, and cold or dry weather retards it. Hemp rets rapidly if spread during early fall, provided there are rains, but it is likely to be less uniform than if retted during the colder months. It should not be spread early enough to be exposed to the sun in hot, dry weather. Alternate freezing and thawing or light snows melting on the hemp give most desirable results in retting. Slender stalks one-fourth inch in diameter or less ret more slowly than coarse stalks, and such stalks are usually not overretted if left on the ground all winter. Hemp rets well in young wheat or rye, which hold the moisture about the stalks. In Kentucky most of the hemp is spread during December. A protracted January thaw with comparatively warm rainy weather occasionally results in overretting. While this does not destroy the crop, it weakens the fiber and causes much loss. When retted sufficiently, so that the fiber can be easily separated from the hurds, or woody portion, the stalks are raked up and set up in shocks, care being exercised to keep them straight and with the butts even. They are not bound in bundles, but a band is sometimes put around the shock near the top. The work of taking up the stalks after retting is usually done by piecework at the rate of $1 per acre.

WATER RETTING

Water retting is practiced in Italy, France, Belgium, Germany, Japan, and China, and in some localities in Russia. It consists in immersing the hemp stalks in water in streams, ponds, or artificial tanks. In Italy, where the whitest and softest hemp fiber is produced, the stalks are placed in tanks of soft water for a few days, then taken out and dried, and returned to the tanks for a second retting. Usually the stalks remain in the water first about eight days and the second time a little longer. In either dew retting or water retting the process is complete when the bark, including the fiber, readily separates from the stalks. The solution of the gums is accomplished chiefly by certain bacteria. If the retting process is allowed to go too far, other bacteria attack the fiber. The development of these different bacteria depends to a large extent upon the temperature. Processes have been devised for placing pure cultures of specific bacteria in the retting tanks and then keeping the temperature and air supply at the best for their development. (Rossi, Giacomo. Macerazione della Canapa. Annall della Regia Scuola Superiore di Agricultura di Portici, s.2, v. 7, p. 1-148, 1907.) These methods, which seem to give promise of success, have not been adopted in commercial work.

CHEMICAL RETTING

Many processes for retting or for combined retting and bleaching with chemicals have been devised, but none of them have given sufficiently good results to warrant their introduction on a commercial scale. In most of the chemical retting processes it has been found difficult to secure a soft, lustrous *fiber, like that produced by dew or water retting, or completely to remove the chemicals so that the fiber will not continue to deteriorate owing to their injurious action. One of the most serious difficulties in hemp cultivation at the present time is the lack of a satisfactory method of retting that may be relied upon to give uniform results without injury to the fiber. An excellent crop of hemp stalks, capable of yielding more than $50 worth of fiber per acre, may be practically ruined by unsuitable weather conditions while retting. Water retting, although less dependent on weather conditions than dew retting, has not thus far given profitable results in this country. The nearest approach to commercial success with water retting in recent years in America was attained in 1906 at Northfield, Minn., where, after several years of experimental work, good fiber, similar to Italian hemp in quality, was produced from hemp retted in water in large cement tanks. The water was kept in circulation and at the desired temperature by a modification of the Deswarte-Lopppens system.*

STEAMING

In Japan, where some of the best hemp fiber is produced, three methods of retting are employed---dew retting, water retting, and steaming, the last giving the best results. Bundles of hemp stalks are first immersed in water one or two days to become thoroughly wet. They are then secured vertically in a long conical box open at the bottom and top. The box thus filled with wet stalks is raised by means of a derrick and swung over a pile of heated stones on which water is dashed to produce steam. Steaming about three hours is sufficient. The fiber is then stripped off by hand and scraped, to remove the outer bark. The fiber thus prepared is very strong, but less flexible than that prepared by dew retting or water retting.

BREAKING

Breaking is a process by means of which the inner, woody shell is broken in pieces and removed, leaving the clean, long, straight fiber. Strictly speaking, the breaking process merely breaks in pieces the woody portions, while their removal is a second operation properly called scutching. In Italy and in some other parts of Europe the stalks are broken by one machine, or device, and afterwards scutched by another. In this country the two are usually combined in one operation.

HAND BRAKES

Hand brakes (Pl. XLVI, fig. 1.), with little change or modification, have been in use for many generations, and even yet more than three-fourths of the hemp fiber produced in Kentucky is broken out on the hand brake. This simple device consists of three boards about 5 feet long set edgewise, wider apart at one end than the other and with the upper edges somewhat sharpened. Above this a framework, with two boards sharpened on the lower edges, is hinged near the wide end of the lower frame, so that when worked up and down by *means of the handle along the back these upper boards pass midway in the spaces between the lower ones. A carpenter or wagon maker can easily make one of these hand brakes, and they are sold in Kentucky for about $5.*

The operator takes an armful of hemp under his left arm, places the butts across the wide end of the brake near the hinged upper part, which is raised with his right hand, and crunches the upper part down, breaking the stalks. This operation is repeated several times, moving the stalks along toward the narrow end so as to break the shorter pieces, and when the hemp appears pretty well broken the operator takes the armful in both hands and whips it across the brake to remove the loosened hurds. He then reverses the bundle and breaks the tops and cleans the fiber in the same manner. The usual charge for breaking hemp on the hand brake in this manner is 1 cent to 1 1/2 cents per pound. There are records of 400 pounds being broken by one man in a day, but the average day's work, counting six days in a week, is rarely more than 75 pounds. In a good crop, therefore, it would require 10 to 15 days for one man to break an acre of hemp. The work requires skill; strength, and endurance, and for many years there has been increasing difficulty in securing laborers for it. It is plainly evident that the hemp industry can not increase in this country unless some method is used for preparing the fiber requiring less hand labor than the hand brake.

MACHINE BRAKES

Several years ago a brake was built at Rantoul, Ill., for breaking and cleaning the fiber rapidly, but producing tow or tangled fiber instead of clean, straight, line fiber, such as is obtained by the hand brake. This machine consisted essentially of a series of fluted rollers followed by a series of beating wheels. Machines designed after this type, but improved in many respects, have been in use several years at Havelock, Nebr., and first at Gridley, then at Courtland and Rio Vista, Cal. These machines have sufficient capacity and are operated at comparatively small cost, the hurds furnishing more than sufficient fuel for the steam power required, but the condition of the fiber produced is not satisfactory for high class twines and it commands a lower price than clean, long, straight fiber.

The Sanford-Mallory flax brake, consisting essentially of five fluted rollers with an interrupted motion, producing a rubbing effect, has been used to a limited extent for breaking hemp. This machine, as ordinarily made for breaking flax, is too light and its capacity is insufficient for the work of breaking hemp.

A portable machine brake (Pl. XLVI, fig. 4) has been used successfully in Kentucky during the past two years. It has a series of crushing and breaking rollers, beating and scutching devices, and a novel application of suction to aid in separating hurds and tow. The stalks are fed endwise. The long fiber,

215

scutched and clean, leaves the machine at one point, the tow, nearly clean, at another, and the hurds, entirely free from fiber, at another. It has a capacity of about 1 ton of clean fiber per day.

Another portable machine brake has been in use in California during the past two years, chiefly breaking hemp that has been thoroughly air dried but not retted. This hemp grown with irrigation, becomes dry enough in that arid climate to break well, but this method is not practicable in humid climates without artificial drying. The stalks, fed endwise, pass first through a series of fluted or grooved rollers and then through a pair of beating wheels, removing most of the hurds, and the fiber, passing between three pairs of moving scutching aprons, each pair followed by rollers, finally leaves the machine in a kind of continuous lap folded back and forth in the baling box.

A larger machine (Pl. XLVI, fig. 3), having the greatest capacity and turning out the cleanest and most uniform fiber of any of the brakes thus far brought out, has been used to a limited extent during the past eight years in Kentucky, California, Indiana, and Wisconsin. This machine weighs about 7 tons, but it is mounted on wheels and is drawn about by a traction farm engine, which also furnishes power for operating it. The stalks are fed sidewise in a continuous layer 1 to 3 inches thick, and carried along so that the ends, forced through slits, are broken and scutched simultaneously by converging revolving cylinders about 12 and 16 feet long. One cylinder, extending beyond the end of the other, cleans the middle portion of the stalks, the grasping mechanism carrying them forward being shifted to the fiber cleaned by the shorter cylinder. The cylinders break the stalks and scutch the fiber on the under side of the layer as it is carried along, and the loosened hurds on the upper side are scutched by two large beating wheels just as it leaves the machine. The fiber leaves the machine sidewise, thoroughly cleaned and ready to be twisted into heads and packed in bales. This machine with a full crew of 15 men, including men to haul stalks from the field and others to tie up the fiber for baling, has a capacity of 1,000 pounds of clean, straight fiber of good hemp per hour. The tow is thrown out with the hurds, and until recent improvements it has produced too large a percentage of tow. It does good work with hemp retted somewhat less than is necessary for the hand brake, and it turns out more uniform and cleaner fiber. For good work it requires, as do all the machines and also the hand brakes, that the hemp stalks be dry. If the atmosphere is dry at the time of breaking, the hemp may be broken directly from the shocks in the field, but in regions with a moist atmosphere, or with much rainy weather, it would be best to store the stalks in sheds or under cover, and with a stationary plant it might be economical to dr them artificially, using the hurds for fuel. Extreme care must be exercised in artificial drying, however, to avoid injury to the fiber.

IMPROVEMENT NEEDED IN HEMP-BREAKING MACHINES

While hemp-breaking machines have now reached a degree of perfection at which they are successfully replacing the hand brakes, as the thrashing machines half a century ago began replacing the flail, there is still room for improvement. This needed improvement may be expected as soon as hemp is grown more extensively, so as to make a sufficient demand for machinery to induce manufacturers to invest capital in this line. For small and scattered crops a comparatively light, portable machine is desirable, requiring not more than 10 horsepower and not more than four or five laborers of average skill for its operation. It should prepare the fiber clean and straight, ready to be tied in hanks for baling, and should have a capacity of at least 1,000 pounds of clean fiber per day. For localities where hemp is grown more abundantly, so as to furnish a large supply of stalks within short hauling distance, a larger machine operated in a stationary central plant by a crew of men trained to their respective duties, like workers in a textile mill, will doubtless be found more economical. Artificial retting and drying may also be used to good advantage in a central plant. The hemp growers of Europe have adopted machine brakes more readily than the farmers in this country, and the hemp industry in Europe is most flourishing and most profitable where the machines are used. Most of the hemp in northern Italy is broken and scutched by portable machines. Machines are also used in Hungary, and the machine-scutched hemp of Hungary is regularly quoted at $10 to $15 per ton higher than that prepared by hand. These European machines may not be adapted to American conditions, but, together with American machines which are doing successful work, they sufficiently contradict the frequent assertion of hemp growers and dealers that "no machine can ever equal the hand brake."

SORTING

On many hemp plantations the stalks are roughly sorted before breaking, so that the longer or better fiber will be kept separate. The work of sorting can usually be done best at this point, short stalks from one portion of a field being kept separate from the longer stalks of another portion and overretted stalks from stalks with stronger fiber. Sometimes the men breaking the hemp sort the fiber as it is broken. An expert handler of fiber will readily sort it into two or three grades by feeling of it as it leaves the hand brake or the breaking machine. It is a mistaken policy to suppose that the average price will be higher if poor fiber is mixed with good. It may be safely assumed that the purchaser fixing the price will pay for a mixed lot a rate more nearly the value of the lowest in the mixture, and he can not justly do otherwise, for the fiber must be sorted later if it is to be used to the best advantage in the course of manufacture.

PACKING FIBER FOR LOCAL MARKET

The long, straight fiber is put up in bundles, or heads, 4 to 6 inches in diameter and weighing 2 to 4 pounds. (Pl. XL, fig.4.) The bundle of fiber is twisted and bent over, forming a head about one-third below the top end. It is fastened in this form by a few strands of the fiber itself, wound tightly around the neck and tucked in so that it may be readily unfastened without cutting or becoming tangled. Three ropes, each about 15 feet long, twisted by hand from the hemp tow, are stretched on the ground about 15 inches apart. The hanks of fiber are piled crosswise on these ropes with the heads of the successive tiers alternating with the loose ends, which are tucked in so as not to become tangled. When the bundle thus built up is about 30 inches in diameter, the ropes are drawn up tightly by two men and tied. These bundles weigh about 200 pounds each. Most of the hemp leaves the farm in this form. Hemp tow, produced from broken or tangled stalks and fiber beaten out in cleaning the long straight hemp, is packed into handmade bales in the same manner.

HACKLING

In Kentucky, most of the hemp is sold by the farmers to the local dealers or hemp merchants. The hemp dealers have large warehouses where the fiber is stored, sorted, hackled, and baled. The work of hackling is rarely done on the farms. The rough hemp is first sorted by an expert, who determines which is best suited for the different grades to be produced. A quantity of this rough fiber, usually 112 or 224 pounds, is weighed out to a workman, who hackles it by hand, one head at a time. The head is first unfastened and the fiber shaken out to its full length. It is then combed out by drawing it across a coarse hackle, beginning near the top end and working successively toward the center. When combed a little beyond the center, the bundle of fiber is reversed and the butt end hackled in the same manner. The coarse hackle first used consists of three or four rows of upright steel pins about 7 inches long, one-fourth of an inch thick, and 1 inch apart. The long fiber combed out straight on this hackle is called "single-dressed hemp." This may afterwards be treated in much the same manner on a smaller hackle with finer and sharper needles set closer together, splitting and subdividing the fibers as well as combing them out more smoothly. The fiber thus prepared is called "double-dressed hemp," and it commands the highest price of any hemp fiber on the American market.

The work of hackling is paid for at a certain rate per pound for the amount of dressed fiber produced. The workman therefore tries to hackle and dress the fiber in such a manner as to produce the greatest possible amount of dressed fiber and least amount of tow and waste. The dressed fiber is carefully inspected before payment is made, and there are few complaints from manufacturers that American dressed hemp is not up to the standard.

A large proportion of the hemp purchased by the local dealers is sold directly to the twine and cordage mills without hackling or other handling except carefully sorting and packing into bales.

BALING

The bales packed for shipment are usually about 4 by 3 by 2 feet. The following table gives the approximate weights per bale:

Average weight per bale of hemp for shipment to mills.

CLASS OF HEMP	POUNDS
Tow	450
Rough	500
Single dressed	800
Double dressed	900

When cleaned by machine brakes the fiber is often baled directly without packing it in the preliminary handmade bales. In this way it has sometimes escaped the process of careful sorting and has brought unjust criticism on the machines. This cause for criticism may easily be avoided by exercising a little more care in sorting the stalks, and, if necessary, the cleaned fiber.

YIELD

The yield of hemp fiber ranges from 400 to 2,500 pounds per acre. The average yield under good conditions is about 1,000 pounds per acre, of which about three-fourths are line fiber and one-fourth is tow. The yield per acre at different stages of preparation may be stated as follows:

Stalks

Green, freshly cut.	15,000 pounds
Dry, as cured in shock	10,000 pounds
Dry, after dew retting	6,000 pounds
Long fiber, rough hemp	750 pounds
Tow	250 pounds

If the 750 pounds of long fiber is hackled it will yield about 340 pounds of single-dressed hemp, 180 pounds shorts, 140 pounds fine tow, and 90 pounds hurds and waste. The average yields in the principal hemp-producing countries of Europe, based on statements of annual average yields for 5 to 10 years, are as follows:

Russia	358 pounds
Hungary	504 pounds
Italy	622 pounds
France	662 pounds

The yield is generally higher in both Europe and the United States in regions where machine brakes are used, but this is due, in part at least, to the better crops, for machine brakes usually accompany better farming.

COST OF HEMP-FIBER PRODUCTION

The operations for raising a crop of hemp are essentially the same as those for raising a crop of wheat or oats up to the time of harvest, and the implements or tools required are merely a plow, disk, drill or seeder, a harrow, and a roller, such as may be found on any well-equipped farm. Estimates of the cost of these operations may therefore be based upon the cost of similar work for other crops with which all farmers are familiar. But the operations of harvesting, retting, breaking, and baling are very different from those for other farm crops in this country. The actual cost will, of course, vary with the varying conditions on different farms. Hemp can not be economically grown in areas of less than 50 acres in any one locality so as to warrant the use of machinery for harvesting and breaking. The following general estimate is therefore given for what may be considered the smallest practical area:

> Estimated cost and returns for 50 acres of hemp
> Cost: Plowing (in fall) 50 acres, $2 per acre...$100
> Disking (in spring), 50 cents per acre.....$25
> Harrowing, 30 cents per acre......$15
> Seed, 40 bushels, delivered, $4.50 per bushel.....$180
> Seeding, 40 cents per acre.....$20
> Rolling, 30 cents per acre.......$15
> Self-rake reaper for harvesting.....$75
> Cutting with reaper, $1 per acre.....$50
> Picking up from gavels and shocking, $1 per acre.....$50
> Spreading for retting, $1.50 per acre.....$75
> Picking up from retting swath and setting in shocks, $1.40 per acre...$70
> Breaking 50,000 pounds fiber, including use of machine brake,
>   1 1/2 cents per pound.....$750
> Baling 125 bales (400 pounds each), including use of baling press,
> $1.40 per bale.....$175
> Marketing and miscellaneous expenses.....$150
> Total cost..... $1,750
> Returns: Long fiber, 37,500 pounds, 6 cents per pound.....$2,250
> Tow, 12,500 pounds, 4 cents per pound.................$ 500
> Total returns.........$2,750

It is not expected that a net profit of $20 per acre, as indicated in the foregoing estimate, may be realized in all cases, but the figures given are regarded as conservative where all conditions are favorable.

MARKET

All of the hemp produced in this country is used in American spinning mills, and it is not sufficient to supply one-half of the demand. The importations have been increasing slightly during the past 20 years, while there has been a decided increase in values. The average declared value of imported hemp,

220

including all grades, for the 4,817 tons imported in 1893, was $142.31 per ton, while in the fiscal year 1913 the importations amounted to 7,663 tons with an average declared value of $193.67 per ton. There have been some fluctuations in quotations, but the general tendency of prices of both imported and American hemp has been upward. (Fig. 19.) The quotations for Kentucky rough prime, since October, 1912, have been the highest recorded for this standard grade. Furthermore, the increasing demand for this fiber, together with the scarcity of competing fibers in the world's markets, indicates a continuation of prices at high levels.

EFFECT OF TARIFF

So far as can be determined from records of importations and prices since 1880, the earliest available statistics, the changes in the rate of import duty on hemp have had no appreciable effect on the quantity imported, on the declared import value ((Declared value of port of shipment.) of the fiber, or on the quantity produced or the price of American hemp in this country. (Fig. 20.) The tariff acts of 1870, 1883, and 1890, in force until 1894, imposed a duty of $25 per ton on line hemp. From 1894 to 1899 hemp was on the free list, and from 1899 to 1913 it was dutiable at $22.50 per ton. The importations reached a high level in 1899, when hemp was extensively used for binder twine. From that year onward henequen from Yucatan and abaca from the Philippines replaced hemp in binder twine, while jute from India replaced it completely for cotton-bale covering. The increasing demand for hemp for commercial twines has resulted in higher prices for both imported and American hemps, but this demand has been met in this country neither by importation nor by production. There are no accurate statistics of acreage or production in the United States, but there has been a general decline from about 7,000 tons in 1880 to about 5,000 in 1913. The average annual production during the period of free importations, 1894 to 1899, was about 5,000 tons, but slightly less than that of the previous 10 years and about the same as the average of the period of dutiable hemp since then. The present tariff, 1913, with hemp on the free list, has not been in force long enough to indicate any appreciable effect.

LOCATION OF AMERICAN MILLS

Some hemp from the larger farms is sold directly to the spinning mills, but most of that produced in this country passes through the hands of local dealers in Kentucky. The hemp imported is purchased either directly from foreign dealers by the mills or through fiber brokers in New York and Boston. (Insert fig. 21 here)

There is one twine mill at Frankfort, Ky., on the western edge of the hemp -producing region, and one at Covington, Ky., opposite Cincinnati, but aside from the comparatively small quantities used by these mills and a little used in the mill at Oakland, Cal., practically all the hemp fiber is shipped away from the States where it is produced. There are 28 mills in this country using American hemp, most of them in the vicinity of Boston or New York, as indicated on the accompanying map (Some of the mills are so close together around New York and Boston that it is impossible to indicate each one by a separate star.) (fig. 21). In most of these mills other soft fibers, such as jute, China jute, and flax, are also used, and many of them are also engaged in the manufacture of twines and cordage from the hard fibers---sisal, henequen, abaca (manila), phormium, and Mauritius.

USES

Hemp is used in the manufacture of tying twine, carpet warp, seine twine, sails, standing rigging, and heaving lines for ships, and for packing. It has been used to some extent for binder twine, but at the relative prices usually prevailing it can not well compete with sisal and abaca for this purpose. Binder twine made of American hemp and India jute mixed has been placed upon the market. This twine is said to give excellent results because it is more smooth and uniform than twine made of hard fiber. The hemp fiber is tougher and more pliable than hard fibers, and the twine is therefore more difficult to cut in the knotter. Hemp is also used to a limited extent for bagging and cotton baling. Only the tow and cheaper grades of the fiber can compete with other fibers for these purposes. The softer grades of hemp tow are extensively used for oakum and packing in pumps, engines, and similar machinery. It endures heat, moisture, and friction with less injury than other fibers, except flax, used for these purposes. Hemp is especially adapted by its strength and durability for the manufacture of carpet warp, hall rugs, aisle runners, tarpaulins, sails, upholstery webbing, belt webbing, and for all purposes in textile articles where strength, durability, and flexibility are desired. Hemp will make fabrics stronger and more durable than cotton or woolen fabrics of the same weight, but owing to its coarser texture it is not well suited for clothing and for many articles commonly made of cotton and wool.

COMPETING FIBERS

The principal fibers now competing with American-grown hemp are Russian and Hungarian hemp, cotton, and jute. Italian hemp, being water retted, is not only higher in price but it is different in character from the American dew-retted hemp, and it is used for certain kinds of twines and the finer grades of carpet warp for which American hemp is not well suited. Twine made of Italian hemp may, of course, be used sometimes where American hemp twine might serve just as well, but owing to its higher price it is not likely to be used as a substitute, and it can not compete to the disadvantage of American hemp. Russian and Hungarian hemp, chiefly dew retted, is of the same character as American hemp and is used for the same purposes. Russian hemp is delivered at the mills in this country at prices but little above those of rough hemp from Kentucky. Most of the Russian and Hungarian hemp imported is of the better grades, the poorer grades being retained in Europe, where many articles are made of low-grade hemp that would be made of low-grade cotton in this country. In some years, owing to unsuitable weather conditions for retting Kentucky hemp or to greater care in handling Russian hemp and to care in grading the hemp for export from Russia, much of the Russian hemp of the better grades has been stronger and more satisfactory to twine manufacturers than American hemp placed on the market at approximately the same price. It is used for mixing with overretted and weak American hemp to give the requisite strength to twine.

Cotton is now used more extensively than all other vegetable fibers combined. The world's supply of cotton is estimated in round numbers at 5,500,000 tons, valued at nearly $1,000,000,000. The total supply of all other fibers of commerce---hemp, flax, jute, China jute, ramie, sisal, abaca, phormium, Mauritius fiber, cabuya, mescal fiber, and Philippine maguey---amounts annually to about 3,300,000 tons, valued at about $350,000,000. Cotton, therefore, so greatly overshadows all other textile fibers that it may scarcely be regarded as competing directly with any one of them. Cotton is prepared and spun on different kinds of machines from those used for preparing and spinning long fibers. Cotton is not mixed with hemp and is rarely spun in the same mills where hemp is used. Cotton twines do, however, compete with hemp tying twines, and cotton is largely used for carpet warp, where hemp, with its superior strength and durability, would give better service. Less than a century ago hemp and flax were used more extensively than cotton, but the introduction of the cotton gin, followed by the rapid development of machinery all along the line for preparing and spinning cotton fiber, while there has been no corresponding development of machinery for preparing and spinning hemp or other long fibers, has given cotton the supremacy among vegetable fibers. There is little probability that hemp will regain the supremacy over cotton, even with improved machinery for handling the crop and spinning the fiber, because cotton is better adapted to a wide range of textile products. Hemp should, however, regain many of the lines where it will give better service than cotton.

Jute is the most dangerous competitor of hemp. Jute is produced in India from the bast or inner bark of two closely related species of plants, jute (Corchorus capsularis) and nalta jute (Corchorus olitorius). These plants are somewhat similar in appearance to hemp, though not at all related to it. They are grown on the alluvial soils in the province of Bengal, India, and to a much less extent in other parts of India, southern China, and Taiwan (Formosa). More than 3,000,000 acres are devoted to this crop, and the annual production is approximately 2,000,000 tons of fiber, valued at $150,000,000. The plants are pulled by hand, water retted in slow streams or stagnant pools, and the fiber cleaned by hand without the aid of even crude appliances as effective as the hand brake for hemp. Jute fiber thus prepared, cleaner, softer, and more easily spun than Kentucky rough-prime hemp, is delivered in New York at an average price of about 4 cents per pound for the better grades. Jute butts, consisting of the coarser fiber cut off at the base, 5 to 10 inches long, are sold in this country at 1 to 2 cents per pound. Most of the long jute fiber comprising the "light jute" grades are of a light straw color, while the "dark jutes," also called "desi jute," are of a dark, brownish gray. The fresh fiber of both kinds when well prepared is lustrous, but with age it changes to a dingy, brownish yellow.

Fresh jute fiber is about two-thirds as strong as hemp fiber of the same weight, but jute lacks durability and rapidly loses its strength even in dry air, while if exposed to moisture it quickly goes to pieces. It is not suitable for any purpose where strength or durability is required. Jute is used most extensively for burlaps, gunny bags, sugar sacks, grain sacks, wool sacking, and covering for cotton bales. Hemp has been used for all of these purposes, but the cheaper jute fiber now practically holds the entire field in the manufacture of coverings for agricultural products in transit. This is a legitimate field for jute, where it constitutes a "gift package," generally to be used but once, but even in this field hemp may regain some of its uses when it is found that jute does not give sufficient strength or durability.

Jute is often used as an adulterant or as a substitute for hemp in the manufacture of twines, webbing, carpet warp, and carpets. The careless use of the name hemp to indicate jute aids in facilitating this substitution. Twine made of pure jute fiber is sold as "hemp twine" in the retail stores in Lexington, Ky., in the heart of the hemp-growing region. Many of the so-called hemp carpets and hemp rugs are made only of jute, and they wear out quickly, whereas a carpet made of hemp should be as durable as one made of wool. Jute is substituted for hemp very largely in the manufacture of warp for carpets and rugs, a purpose for which its lack of strength and durability makes it poorly fitted. It is to the interest of the purchaser of manufactured articles as well as to the producer of hemp and the manufacturer of pure hemp goods that the line between hemp and jute be sharply drawn. Unfortunately, the difference in the appearance of the fibers by which they may be distinguished is not as strongly marked as the differences between their strength and wearing qualities.

TESTS FOR DISTINGUISHING BETWEEN JUTE AND HEMP

There are no satisfactory tests for these fibers without the aid of a microscope and chemical reagents. A ready, but uncertain, test consists in untwisting the end of twine or yarn. Jute fiber thus unwound is more fuzzy and more brittle than hemp. The two fibers may be distinguished with certainty with a microscope and chemical reagents, as indicated by the differences in the table which follows:

REACTIONS OF HEMP AND JUTE

At the present high prices of jute (fig.4), resulting from increasing demands in foreign markets and a partial failure of the crop in India, jute could not compete successfully with hemp were it not that manufacturers are using it in established lines of goods, and, further, that they are uncertain about securing supplies of hemp.

JUTE FIBER DRYING IN THE SUN
ABUTILON AVICINNOE. [CORCHORUS SP.]

AMERICAN JUTE. NEAR YANG TSUN, CHINA. VIEW OF ONE SIDE OF A PIECE OF A NICE RICK OF GOOD LOOKING "CHING MA" FIBER DRYING IN THE SUN. THIS "CHING MA" FIBER PLANT IS SAID TO PRODUCE 80 CATHIS, ABOUT 100 POUNDS, OF FIBER PER ACRE. THE LONG FIBER SELLS AT 14.00 LOCAL DOLLARS PER 120 CATHIS AND 13.00 PER THE SHORT FIBER.PHOTOGRAPH #45580 –

WWW.NAL.USDA.GOV/.../DORSETT/ MISCPICS.HTML

SUMMARY

Hemp is one of the oldest fiber-producing crops and was formerly the most important.

The cultivation of hemp is declining in the United States because of the (1) increasing difficulty in securing sufficient labor for handling the crop with present methods, (2) lack of labor-saving machinery as compared with machinery for handling other crops, (3) increasing profits in other crops, (4) competition of other fibers, especially jute, and (5) lack of knowledge of the crop outside of a limited area in Kentucky.

Hemp was cultivated for fiber in very early times in China.

The history of the distribution of hemp from Asia to other continents indicates its relationships and the development of the best fiber-producing types.

Hemp is cultivated in warm countries for the production of a narcotic drug, but for fiber only in moderately cool and humid temperate regions.

Very few well-marked varieties of hemp of fiber-producing types have been developed.

The climate and soils over large areas in the valley of the Mississippi and its tributaries and in the Sacramento and San Joaquin Valleys in California are suited for hemp.

Hemp improves the physical condition of the soil, destroys weeds, and when retted on the ground, as is the common practice, does not exhaust fertility.

Hemp is recommended for cultivation in regular crop rotations to take the place of a spring-sown grain crop.

Fertilizers are not generally used in growing hemp, but barnyard manure applied to previous crops is recommended.

Hemp is rarely injured by insects or fungous diseases.

Broom rape, a root parasite, is the most serious pest in hemp.

Practically all of the hemp seed used in the United States is produced in Kentucky.

The best seed is obtained from plants cultivated especially for seed production, but some seed is obtained from broadcast overripe fiber crops.

The land should be well plowed and harrowed, so as to be level and uniform.

The seed should be sown early in spring by any method that will distribute and cover it uniformly.

Some hemp is still cut by hand in Kentucky, but the use of machinery for harvesting the crop is increasing.

Dew retting is regarded as the most practical method in this country. Hand brakes for preparing the fiber are still used, but they are being replaced by machines.

The price of hemp has been generally increasing during the past 30 years.

About 30 different spinning mills in the United States, beside dealers in oakum supplies, offer a market for raw hemp fiber.

The market would expand if manufacturers could be assured of larger supplies.

Source: www.naihc.org/hemp_information/content/1913.html

HEMP-KNOWLEDGMENTS

This book started as a response to the crisis which our planet home finds itself in 2007.

The impetus for the book came from our editor
J. Nayer Hardin
who worked tirelessly to get our message to the public, industry, government and the world at large. She deserves special thanks for getting me started on this project.

Thanks to Nayer's partner,
Sherwood Akuna
who added artwork and design

Special thanks to my companion,
Brenda Kershenbaum
who joined with me in 1995
to form the World Cannabis Foundation.
She has made many contributions to this book including research, and the gathering of museum artifacts, for which we both hold the vision of an actual Hemp Museum in Los Angeles, where people will be able to see the wonder of hemp first-hand. Brenda also did initial financing and promotion of the 1990 edition of The Emperor Wears No Clothes,
the book which inspired me
to create the USA Hemp Museum.

Thanks goes to all the unnamed hemp activists who would not let the idea die that hemp could save the planet.

227

THE RESEARCH APPLICATION

Los Angeles, CA - The USA Hemp Museum's Founder and Curator, Richard M. Davis announced that a HEMP FOR VICTORY plan is in development at the museum to successfully use hemp to help solve the survival problem of global warming.

Why hemp? Hemp dynamically reduces pollution in two major ways. First, hemp is biomass champion, breathing in more carbon dioxide (the most abundant greenhouse gas) than any other plant. This carbon dioxide is turned into wood and fiber by photosynthesis. Hemp wood takes the pressure off our forests by making paper and building materials like pressboard. Also, Hemp can do all the jobs fossil fuels (coal, oil, and natural gas) do now. When used as bio fuel, hemp replaces toxic energy (fossil fuels, nuclear power) with clean sustainable energy. Hemp biofuel can be processed to run any engine, heat or cool any building, run any factory, and eliminate the greenhouse gases and pollution that come from modern energy sources. Remove the cause, pollution, and the effect, global warming can be reduced, if not healed.

The HEMP FOR VICTORY plan is considering the problem of global warming, benefits of hemp, land use and availability, financing with a hemp tax, restoring the family farm to grow hemp, seeds, irrigation, growing techniques, harvesting, processing, replacing toxic energy, how to use existing energy distribution systems, and other benefits from re-hemping the earth. This plan is based on the HEMP FOR VICTORY campaign of World War II (see USA Hemp Museum's History Room).

"A state of international emergency has been declared by many of the top scientists in the world." Davis said "We have a solution. To help save ourselves from the ravages of global warming, we must grow massive quantities of hemp all over the world to reduce carbon dioxide and provide bio fuel. Working together, crops can be bred for maximum earth healing affect. Markets can be coordinated. The impact of restoring the family farmers can help those who are not currently benefiting from our alleged booming economy by keeping energy dollars at home.

Decisions have consequences and our decision to stay on toxic energy is seriously impacting global warming. We must change now to survive and hemp is key to the process."

For more information, to help establish an immediate seed fund, build the physical hemp museum, donate, or to contribute hemp items *and stories contact Richard Davis, The USA Hemp Museum, www.hempmuseum.org.*

**RE-HEMP THE PLANET-
CLEAN THE ENVIRONMENT.
DO IT NOW.**

**UNDERSTAND THE PROBLEM
AND IMPLEMENT
EFFECTIVE SOLUTIONS**

STOP POLLUTING ACTIVITIES

**INSTITUTE A 20% RECREATIONAL HEMP TAX
TO COVER THE COST**

**DEVELOP AND IMPLEMENT A
HEMP FOR VICTORY PROGRAM, LIKE IN WWII**

**GROW HEMP EVERYWHERE POSSIBLE
TO CLEAN THE AIR AND ADD OXYGEN**

**MANAGE THE MELTING ICEBERGS FOR IRRIGATION
AND DRINKING WATER**

**PRAY AND MEDITATE TO BECOME
GOOD CARETAKERS OF THE EARTH**

FREE ALL HEMP POLITICAL PRISONERS

TO ORDER ADDITIONAL COPIES:

THE U.S.A. HEMP MUSEUM

WWW.HEMPMUSEUM.ORG

HAPPY HEMP

www.ingramcontent.com/pod-product-compliance
Lightning Source LLC
Chambersburg PA
CBHW050837220326
41598CB00006B/382